A Guide to
Earth Satellites

A Guide to Earth Satellites

edited by

DAVID FISHLOCK

Macdonald : London
AND
American Elsevier Inc. ; New York

© Macdonald & Co. (Publishers) Ltd, 1971

Sole distributors for the United States and Dependencies
American Elsevier Publishing Company, Inc.
52 Vanderbilt Avenue
New York, N.Y. 10017

Sole distributors for the rest of the World
Macdonald & Co. (Publishers) Ltd
P.O. Box 2 L.G.
49–50 Poland Street
London W.1

Macdonald SBN 356 03867 X
American Elsevier ISBN 0 444 19581 5
Library of Congress Catalog Card Number 79 167624

Made and Printed in Great Britain by
Butler & Tanner, Ltd, Frome and London

Contents

Introduction

DAVID FISHLOCK
Financial Times

This book describes satellites in Earth orbit; circumnambulating spacecraft designed to gather and relay information of many different kinds, telephone, TV or digital, for example, navigational and meteorological guidance, or intelligence of scientific, commercial or political significance. Such craft are providing communications of all kinds on Earth with a new dimension, potentially of immense value, the technology of which has barely reached its 'teens'.

Of the eight distinguished contributors, six describe the major areas of application envisaged at present, for 'comsats', 'metsats', natural resource satellites, 'navsats' and satellites for gathering data relating to science, and warfare. The other two authors introduce and conclude the book with chapters discussing, respectively, the technology common to all kinds of Earth satellites, and an ambitious proposal for a power station in space on a scale large enough to harness the Sun economically.

All of the spacecraft discussed in detail in these eight chapters are robots, designed to operate unmanned. Yet the main goal now of the world's two major space research programmes is the manned Earth satellites. The first was the Russian satellite *Salyut*, launched in the Spring of 1971, whose crew died so tragically at the end of their mission. The aim of manned orbiting satellites, however, far from being the displacement of the robot, might be seen as a key contribution to the performance of unmanned spacecraft, initially by using men to decide at first hand what the instruments need to observe, and later by affording the prospect of a repair and maintenance service in space.

To this end, the US has planned *Skylab*, a manned spacecraft, whose relevance to the development of applications satellites is discussed in several of the chapters which follow. The National Aeronautics and Space Administration, having accomplished its first lunar landing in 1969 with fewer spaceflights than were originally planned, was left with an inventory of Saturn 5 rockets, around which it has planned an orbital workshop. In 1973 it plans to place in Earth orbit at 270 miles altitude a Saturn 4B upper-stage fuel tank fitted out as a workshop, to be manned by crews of three who may remain on-station as long as 120 days. A second space workshop is planned for 1974, with an operational life of 12–18 months. Then, in 1975, NASA hopes to assemble its first space station, serving for 5 to 10 years.

Skylab 1 will orbit more than 50 experiments, embracing up to a dozen different disciplines, to be carried out by technically qualified astronauts. Mostly they will relate to the instrumentation required for unmanned spacecraft, notably in the areas of Earth and Space surveillance. But one series of experiments could lead to an entirely new application for spacecraft: space manufacture.

Space manufacture

It is easy to be cynical about NASA's enthusiasm for its 'Manufacturing in Space' programme while launch costs alone remain in the region of $500 to $1500 per lb. But if we accept its forecasts – and the space agency has a good reputation for its cost estimates – the cost will plummet with the introduction of reusable launchers, the so-called 'space shuttle' concept, around the mid-1970s. Launch costs eventually as low as $50 and even $30 per lb have been predicted by Dr Werner von Braun, head of NASA's Marshall Space Flight Center.

Already there are many electronics materials selling in the region of $10 000 to $100 000 per lb, a price range which reflects the immense problems of manufacture on Earth. Minute chips of such crystals form the heart of devices that fetch up to $5000 apiece. Closer to general engineering are the high-modulus fibres such as carbon and boron, which like crystals are dependent on purity and perfection.

Zero-*g* is the focus of interest so far. Apart from two residual sources of acceleration – the spacecraft's own thrusters for station keeping, which need not be used while experiments are in progress; and the astronauts' movements, which would impart accelerations of only thousandths of a *g* – an orbital facility would be zero-*g*. For liquid-state operations it affords fascinating possibilities for purifying, homogenizing and perfecting the structure of both materials and products.

But there is another unique aspect of the space environment that may prove worth harnessing, namely the extremely hard vacuum of space. Even in a low Earth orbit there exists in the wake of the spacecraft a cone of ultra-high vacuum comparable with anything yet achieved on Earth.

Cheap boron filament

Let us look first at a material in which aerospace engineers themselves have shown great interest, boron filament. Developed by the US Air Force as a strong, stiff fibre for composite structures, it has already been used to reinforce experimentally some large aerospace components, including long fan blades, helicopter rotors and horizontal stabilizers. Then along came Britain's carbon fibre, with similar properties but much better prospects for a sharp fall in manufacturing costs as demand rose. For that reason, chiefly, carbon – mostly British made – caught up very quickly and has now overtaken boron in the US aerospace market.

But an investigation by Grumman Aerospace Corporation indicates that manufacture of boron in space might redress the balance in costs and yield a still better fibre.* Grumman's case rests on a new route to boron fibre, impossible on Earth, in which a large ball of molten boron – an extremely corrosive substance with a melting point around 2070°C – would be levitated out of all contact with solids, then forced by hot gas or an induction coil to extrude through an 'electromagnetic die' into a continuous filament.

Such a route might have several advantages to offset the premium on space manufacture. Production rates should be 100–1000 times greater than are possible now, by the process of depositing boron vapour on a tungsten filament. Moreover, by eliminating the tungsten core one might expect to remove the source of most faults in the fibre, and hence improve its strength and stiffness far beyond anything possible at present. Average tensile strengths of well over 1 million lb per square inch are predicted. Moreover, the absence of tungsten – a neutron absorber – would also allow the fibre to be used in nuclear aerospace applications, expected to become a large part, around 25 per cent, of the aerospace market.

Sparklers from space

Crystals of one kind or another can be found at the heart of most new technology, controlling, calculating, recording and communicating the machines' behaviour. The perfection and purity required in their structure if the manufacturer is to secure both performance and yield has forced new crystal-growing techniques quickly from the laboratory into production lines. Yet many potentially highly useful materials remain unexploited. Either they cannot attain the structural perfection required, or they have proved too corrosive when molten to keep pure enough.

* *NASA Report, Space Processing and Manufacturing*, no. ME–69–1, 1969, page 79.

Zero-*g* offers an answer to both of these handicaps. Growth from levitated melts out of all contact with a crucible would prevent contamination. Absence of convection currents would eliminate many, perhaps most, of the flaws from the crystal into which it froze. Of all space manufacture proposals, concludes North American Rockwell, 'it seems to us that crystal-growing offers the most promise of rapid economic success'.*

Gallium arsenide, an intermetallic compound with a bright future as a solid-state microwave (radar) and laser beam source, is a crystal extremely difficult to grow in large sizes, with the perfection these devices demand. Westinghouse Electric has designed an experiment to be flown in *Skylab* 1, in which gallium arsenide crystals 0·5 sq in across will be grown from solution, a technique normally restricted to very small crystals.

Joints in space

Russia's *Soyuz* spacecraft experiments in 1969 with electron-beam welding were the first attempts to join metals in space. A high-precision, high-quality process that can use the very high vacuum of space, electron-beam welding is the most promising method of assembling structures for service in space, and perhaps also for some premium parts to be brought back to Earth. Tolerances for electron-beam welds would be much less critical than down here on Earth, where parts must abut very precisely. Under zero-*g*, cohesion within the molten region would accommodate irregularities, allowing the use of less sophisticated tooling. Dispersion-hardened composite materials such as TD Nickel have never been welded successfully on Earth. Agglommeration and massive segregation of the oxide particles takes place, due respectively to thermal convection and gravity forces, it is believed. For that reason space assembly of such superalloys may prove attractive. Simpler, although restricted to temperatures below 1370°C, is exothermic brazing, in which the heat of a chemical reaction is used to fuse metal. It needs no elaborate apparatus, beyond an igniter and its power supply. Experiments in both electron-beam welding and exothermic brazing are planned for *Skylab* 1.

Alloys unknown

Under weightless conditions it should be possible to form a large number of alloys impossible on Earth. Take, for example, the situation where two metals are totally immiscible when molten. Such

---

* *NASA Report, Space Processing and Manufacturing*, no. ME–69–1, 1969, page 79.

pairs separate so completely that no matter what is done on Earth to homogenize them, they will segregate upon cooling. Remove gravity, however, suggests TRW Systems Group, 'and it may be possible in a variety of immiscible alloy systems to produce homogenized alloys from the melt'.*

Examples offered by the TRW scientists include solid lubricants, such as copper–lead alloys – important bearing materials – containing much more than the 30 per cent of lead to which they are limited now. Another group are directionally solidified composites, where zero-*g* offers new combinations not possible with the eutectics used now, and an extremely uniform distribution of the reinforcing phase. A third group comprises homogenized mixes of metals and non-metals (oxides, carbides, nitrides) to form a new kind of cermet, much stronger than those used today.

Buoyancy rather than immiscibility is the problem with another class of prospective alloys. For instance, small additions of rare earth metals (lanthanum or cerium) to superalloys will greatly reduce their susceptibility to marine corrosion, a serious problem with the blades of gas turbines. Unfortunately, the additions seriously weaken the blade, because the much lighter rare-earth oxide segregates on cooling. Powder techniques are the only way now to sustain a fine dispersion. US General Electric, whose turbine designers want the kind of brine-resistant superalloy zero-*g* manufacture may offer, has worked out techniques for melting such alloys in space.

* *NASA Report, Space Processing and Manufacturing*, no. ME–69–1, 1969, page 383.

CHAPTER **1**

Technology of Earth Satellites

Dr J. A. Vandenkerckhove
European Space Research Organization

For the last decade, Earth satellite systems have been developed and operated successfully to provide a variety of services, primarily in the fields of telecommunication, meteorology and military applications. During the next decade, improved systems will be launched for these applications as well as for new ones, in particular air and sea traffic control and Earth resources. I would not expect, however, the technology of applications satellites to depart greatly from today's technology before the second half of the 1970s. For one thing, the lead time between conception and launch of a typical Earth satellite is between three and six years. More fundamentally, almost all foreseen space applications can be carried out effectively by relatively light, unmanned satellites, just a few hundred kilograms in weight.

It is difficult to foresee how the availability of the large manned space stations under consideration for launch late in the 1970s might influence the future of space applications. True, such stations would permit the grouping of many functions aboard a single spacecraft and ensure the maintenance of the equipment by the astronauts. Conceivably they might also permit us to carry out in orbit some of the tasks normally performed on the ground, such as data processing and the control of Earth-bound activities. Nevertheless, we must keep in mind the high cost of supporting Man in space, the diversity of orbits and positions required to embrace the whole spectrum of space applications, and the relative ease with which it is possible to keep a satellite system under permanent or near-permanent ground control. In this light, it is not clear that the availability of large manned space stations could improve the usefulness or cost-effectiveness of space applications, unless completely new missions, such as space manufacture under weightless and high-vacuum conditions (discussed in the Introduction), come into being.

The launch vehicle

The primary functions of the launch vehicle are to impart to the satellite a very large velocity, of the order of 8000 to 10000 m/s, and to bring it from the pad up to its injection point into orbit. In addition, the launch vehicle often performs secondary functions related to the dynamics of the satellite, such as spin-up or attitude control up to injection.

All launch vehicles used so far have been wingless and expendable multi-stage rockets, although large winged and reusable launch vehicles, often referred to as 'shuttles', are being considered in relation to the manned space stations I have mentioned. The rockets presently in use are launched vertically (or nearly vertically for the smallest types) to minimize the structure needed to support and guide the rocket before it leaves the launcher.

The early part of the flight path is kept nearly vertical in order to minimize drag and aerodynamic heating in the denser layers of the atmosphere. High-thrust rocket motors are always used on the first stages; indeed, by necessity, the total thrust at lift-off must be larger than the total weight of the vehicle. Depending on design, the initial acceleration is between a fraction of a g and a few gs, but it grows rapidly as propellent is expended and the vehicle becomes lighter. Another reason for using high thrusts during the early part of the flight path is that it reduces the gravity losses which result from the need to use propellent to overcome gravity. (A motionless rocket would waste all its propellent overcoming gravity without accelerating.)

In practice, however, the thrust is limited by
(1) the maximum acceleration which occurs near the burn-out of the stage and which should preferably not exceed 10 to 15 g;
(2) the engine deadweight; and
(3) the desirability to keep low aerodynamic drag and heating.
Subsequently, when velocity becomes a significant fraction of the orbital velocity and when the flight path becomes nearly horizontal, centrifugal force becomes important and reduces accordingly the effective gravity force. Near orbital velocity or above it, both high-thrust and low-thrust motors are being used since propulsion is no longer needed to overcome gravity.

Usually each stage of the launch vehicle consists mainly of one or several rocket engines, the propellent tanks, and the payload. The velocity increment Δv which can be imparted to a payload by a single stage is given by the following relationship:

$$\Delta v = kgI_\mathrm{s} \log_\mathrm{e} \frac{M_0}{M_\mathrm{f}}$$

2

where k is a factor which takes into account the losses due to drag and gravity and is approximately between 0·85 and 0·90 for a first stage, 0·92 and 0·96 for an intermediate stage, and 0·96 and 1·00 for an upper stage; g is the acceleration of gravity at sea level (9·81 m/sec²); I_s is the specific impulse of the propellant, which is the reciprocal of the specific consumption and which is always expressed in seconds (it is the number of seconds during which 1 kg of thrust is produced with 1 kg of propellant); M_0 is the initial mass of the stage; and M_f is the final mass of the stage.

The specific impulse of chemical propellents does not exceed about 400 sec, even for the most energetic combination of hydrogen with oxygen or fluorine. The overall mass ratio of an advanced stage does not exceed about 8, corresponding to a propellant fraction of 0·875 and to a rather light payload fraction of approximately 0·025. Hence even with a very advanced technology it is not practical to design a single-stage rocket capable of imparting to a satellite the velocity of 8000 m/sec, required to inject it into a very low orbit.

This limitation in performance of single-stage rockets can be readily explained by the penalty arising from the need to accelerate the whole deadweight up to the final velocity. Indeed, near burn-out, most of the propulsion effort is expended in accelerating almost empty propellant tanks, a high fraction of the deadweight.

The ideal rocket is one able to lose progressively during combustion that fraction of its deadweight which is no longer needed. Such a rocket is not practical but can be approximated through multi-staging. In a multi-stage rocket, the first stage accelerates the upper-stage(s) payload combination to velocity Δv_1. Once its propellent is exhausted it separates. The second stage ignites and accelerates up to velocity $\Delta v_1 + \Delta v_2$ before it too burns out and separates. The process is repeated as many times as there are stages. In theory, such a solution permits the attainment of any required velocity, if at the price of a severe increase in overall weight. Multi-staging is achieved either through series staging, the most popular solution, or through parallel staging, or by a combination of the two.

Fig. 1 represents a cutaway of the Delta vehicle, widely used by the National Aeronautics and Space Administration (NASA) to launch Earth satellites. It consists of two or three stages in series with the possibility to strap in parallel on the first stage from three to nine solid-fuel motors to boost the payload it can orbit. With the chemical propulsion systems available today it is possible to inject a satellite into a low orbit (below 1000 km) with only two stages and an overall payload ratio between 40 and 160. This is the ratio between the total weight of the vehicle at lift-off and the payload weight. For instance, a two-stage Delta with a lift-off weight of

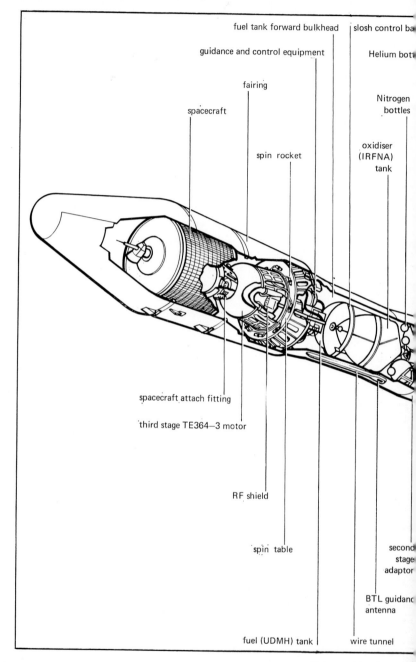

fuel tank forward bulkhead

slosh control ba

guidance and control equipment

Helium bott

fairing

Nitrogen
bottles

spacecraft

oxidiser
(IRFNA)
tank

spin rocket

spacecraft attach fitting

third stage TE364–3 motor

RF shield

spin table

second
stage
adaptor

BTL guidanc
antenna

fuel (UDMH) tank

wire tunnel

FIG. 1   Long-tank De

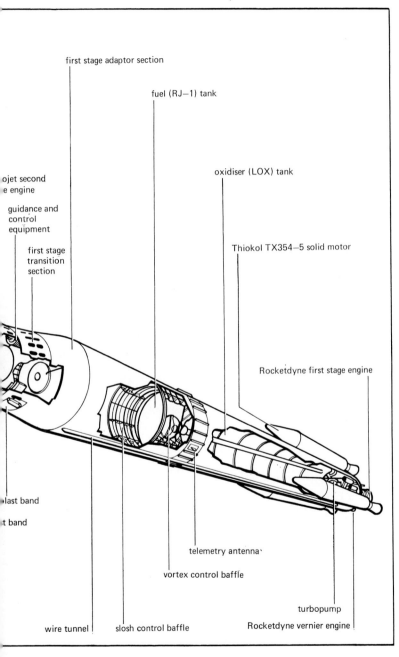

first stage adaptor section

fuel (RJ—1) tank

oxidiser (LOX) tank

ojet second
e engine

guidance and
control
equipment

first stage
transition
section

Thiokol TX354—5 solid motor

Rocketdyne first stage engine

last band

t band

telemetry antenna·

vortex control baffle

turbopump

wire tunnel   slosh control baffle   Rocketdyne vernier engine

pace research vehicle.

about 90 tons, is capable of injecting into a 500-km circular orbit a payload of 1250 kg with three additional rockets, and 1610 kg with nine extra rockets.

A high-performance liquid-hydrogen liquid-oxygen rocket motor for the second stage, instead of the present medium-performance engine which uses kerosene and nitrogen tetroxide as propellent, would increase by a factor $1\cdot5$–2 the payload it can orbit.

In general also, an additional stage is needed when a more demanding orbit is desired. For instance, a three-stage Delta with nine strapped-on rockets is required to place a 600-kg payload consisting of the satellite and its apogee motor (or fourth stage) into a transfer orbit with a perigee at 300 km and an apogee at the geostationary altitude of 35 850 km.

The task of the launch vehicle is not only to impart to the spacecraft its required orbital energy, but also to carry it up to the required injection point. In order to minimize the gravity losses, it is most desirable to ignite the stages in short sequence. However, the burning time of two or three stages is short and it is generally impossible by this means to reach the injection altitude at burn-out of the last stage. It is therefore necessary to introduce before ignition or reignition of the last stage a coasting period; that is, a segment of ballistic trajectory during which the last stage with its spacecraft converts a fraction of its kinetic energy into potential energy in order to arrive at injection altitude when the final kick is provided by the last stage. When the coasting period is short, its energy level is appreciably lower than the orbital energy, and a large impulse is needed from the last stage.

This technique is called the direct ascent into orbit. It is efficient only for the lowest orbits. For higher orbits, it is more efficient to burn the lower stages in short sequence in order to bring the last stage with its spacecraft into a long transfer orbit nearest to the energy level of the final orbit, in which case the required impulse at injection is smaller. This last technique is known as the injection through a Hohmann transfer. Fig. 2 illustrates the two techniques.

A Hohmann transfer is always used when high orbits are required because this improves very significantly the payload capability. For very high orbits, such as the geostationary ones, the transfer lasts for about 5 hours and 45 minutes and the kick at injection is often provided by a separate apogee motor. For instance, out of the 600-kg assembly launched by a three-stage Delta into a transfer orbit to geosynchronous altitude, about 330 kg must be used by a solid-fuel apogee motor fourth stage, needed to circularize the orbit of a 270-kg satellite. Another reason for using segments of orbits, known as parking orbits, before the final propulsive phase(s) is to orient the major axis of either a final or a transfer orbit in the required direction, when this cannot be achieved through direct ascent. The use of a parking orbit is also illustrated by Fig. 2.

Finally, propulsive phases are sometimes required to modify the orbit. For example, we may want to change the plane of the orbit, by a so-called 'dog-leg' manœuvre. This is always a costly manœuvre in terms of propellent, which should preferably be performed at the highest possible altitude where the velocity is smallest. Or we may wish to change the period of the orbit, or the altitude of the apogee

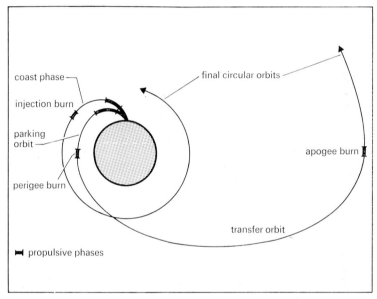

Fig. 2 Two orbit injections techniques: direct ascent for low orbits, and Hohmann transfer for high orbits.

or the perigee. All these changes can be made by imparting to the spacecraft a small velocity increase or decrease at perigee or apogee, respectively.

Constraints on the vehicle

The use of launch vehicles is constrained by many conditions which must be taken into account in system analysis and satellite design:

Geographical restraints: in particular

(1) the latitude of the range, which limits the orbital inclination possible without 'dog-legging' to a value equal to or higher than this latitude; and

(2) the availability of safe impact areas, which often restricts severely the direction in which launch is permitted (in this connection, orbiting of the second stage considerably alleviates the safety problems).

Availability of down-range stations to monitor and possibly tele-command the launch vehicle, and the spacecraft can also, in many instances, impose an additional restraint.

Time restraints: conditions favourable for a launch, such as position of the Sun with respect to the spacecraft and its orbit, are in general only met during short periods of time, called launch windows. Nowadays very short windows, 5 minutes or less, can be accepted.

Vehicle restraints: the vehicle imposes severe restraints on the satellite. First, dimensionally – the satellite must fit into the shroud or heat shield at the top of the rocket. This generally limits severely the design, in particular for heavy payloads, and often we must resort to the use of paddles and booms. Second, environmentally – the satellite must survive the acceleration, vibration and heat. Third, dynamically – use of a solid-propellent last stage requires us to spin-up at a rather high rate, above 100 rev/min, the last stage and satel-

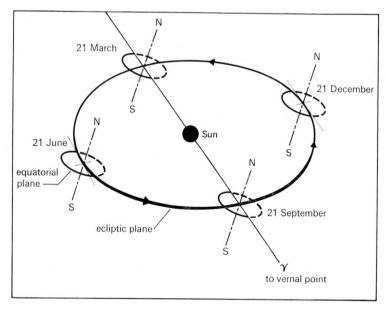

FIG. 3   Earth's motion round the Sun defines the ecliptic plane.

lite assembly in order to neutralize the thrust misalignment. Often later the satellite has to be despun.

Attitude restraints: for a solid-propellent rocket the initial attitude of the satellite is imposed by the direction of the thrust; for a three-axis stabilized, liquid-motor last stage, attitude manœuvres can often be performed before or just after injection. This provides good flexibility for minimizing thermal control problems on board of the satellite, and ensures final optimum attitude.

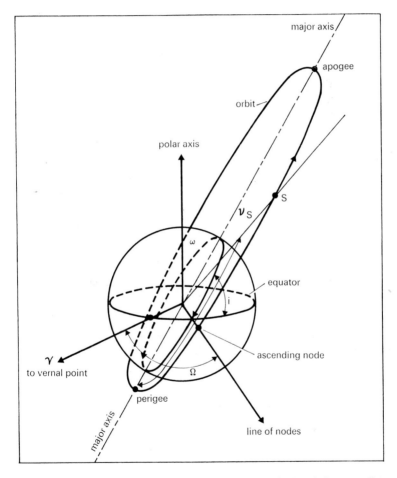

Fig. 4 Orbital plane, defined by inclination and longitude of the ascending node of the orbit.

Choice of orbit

The choice of the satellite orbit has a predominant influence on an Earth application mission. For such missions two categories are of utmost importance:
(1) Low, near-polar orbits
(2) High, geostationary orbits.

The Earth's motion around the Sun defines the ecliptic plane, inclined by 23·4 degrees over the equator (Fig. 3). At the Spring Equinox (21 March) the Sun is both in the equatorial and ecliptic planes and, seen from the Earth, its position corresponds to that of the vernal point of the celestial sphere.

For Earth missions, the orbit is, in general, defined in the equatorial coordinate system, with the following three axes:
(1) The polar axis, positive northwards
(2) The lines of equinoxes, that is the intersection between the equator and the ecliptic planes, positive to the vernal point
(3) The perpendicular to the line of equinoxes, in the equatorial plane.

The orbital plane is defined by the following two angles (see Fig. 4):
(1) The inclination i (the angle between the orbit plane and the equator)

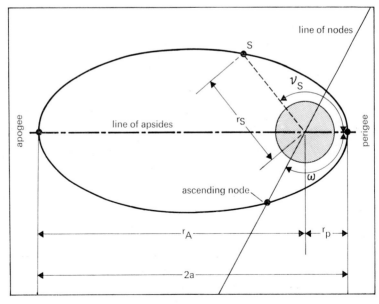

FIG. 5 Unperturbed orbit, approximately a Keplerian ellipse.

(2) The longitude of the ascending node Ω (the angle between the direction of the vernal point and that of the ascending node, seen from the centre of the Earth).

The lines of nodes is the intersection between the orbital and equatorial planes.

To a first approximation the unperturbed orbit of an Earth satellite is a Keplerian ellipse, as represented by Fig. 5, with one focus at the Earth's centre.

Owing to the non-sphericity of the Earth, atmospheric drag, radiation pressure and the gravitational influence of other celestial bodies (mostly the Moon and the Sun), the elliptic motion of the satellite is perturbed. For our present purpose it is sufficient to indicate that these perturbations can be rather significant and must be taken into account in mission studies. As we shall see, it is even possible in certain cases to take advantage of the perturbations.

Low, near-polar orbits

For low-altitude observation of the entire Earth, required for certain meteorological, Earth resources or military missions, we must select a near-polar and usually circular orbit. If a truly polar orbit ($i = 90$ degrees) is selected, the longitude of its ascending node (Ω) will not vary under the effect of the orbital perturbations. Thus the orbital plane will remain fixed in inertial space, and the angle between the orbital plane and the direction of the Sun will vary as the Earth rotates around the Sun. When the orbital plane contains the Sun, the local time at the sub-satellite point is about noon or midnight. Three months afterwards the Sun is perpendicular to the orbital plane, and the local time is about 6.00 a.m. or p.m.

A truly polar orbit therefore permits us to observe the Earth within a three-month period, over the complete range of local times (solar illuminations).

For certain missions, on the other hand, we want to observe the Earth day after day over a long period of time, always at the same local time. This can be done by selecting a near-polar, Sun-synchronous orbit with a longitude of ascending node Ω rotating under the influence of the perturbations due to Earth oblateness at the same rate as the apparent motion of the Sun around the Earth (0·986 degree/day). Sun-synchronization is achieved with an inclination of 98 degrees for a circular orbit at 500 km, and 102 degrees at 1700 km.

Fig. 6 represents, for this latter case, the ground trace of the satellite during a 24-hour period. It also represents the area observed during one of the 12 daily orbits, by a scanner with a field view of 75 degrees pointing to the centre of the Earth. The equatorial region is

11

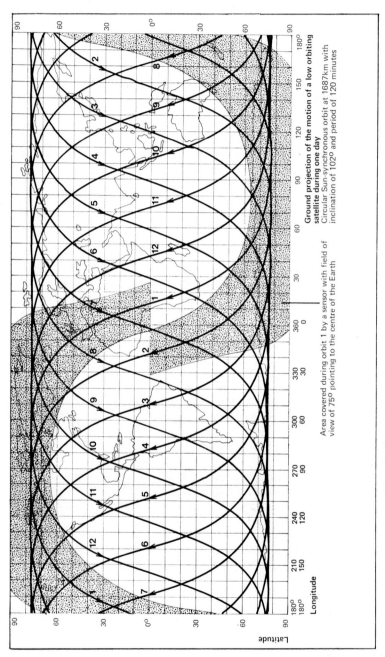

Ground projection of the motion of a low orbiting
satellite during one day
Circular Sun-synchronous orbit at 1687km with
inclination of 102° and period of 120 minutes

Area covered during orbit 1 by a sensor with field of
view of 75° pointing to the centre of the Earth

Fig. 6 Ground trace of a near-polar, Sun-synchronous satellite.

completely observed once every 12 hours; the polar regions once per orbit, with a large overlap at medium latitudes.

Sun-synchronous orbits are used in particular for meteorological satellites (see Chapter 3), to observe clouds at fixed local times. They often simplify satellite design.

High, geostationary orbits

High, geostationary orbits are very often used when permanence of services over a given geographical area is required, for telecommunications for example.

A circular orbit at an altitude of 35 850 km has a period equal to that of the Earth's rotation. In equatorial orbit ($i = 0$ deg.), the satellite will remain constantly over the point on the equator above which its injection has taken place. In practice, the satellite tends to drift slightly, but the perturbations are of relatively small amplitude and can be easily countered by a station-keeping system.

Such a satellite sees almost half the Earth's surface (Fig. 7). For example, one satellite can see simultaneously Europe, the Middle East, Africa, South America and a significant part of North America. Such a satellite is obviously suited for providing translatlantic communications. From such a satellite the Earth is seen within a cone of 18 degrees aperture. Much smaller fields of views can be used to cover smaller areas. Individual European countries such as Britain, France, Germany or Italy would require a beam only about 1 to 2 degrees in aperture.

The geostationary orbit, with its overwhelming advantages, is unique and might well before long become congested over such areas as the Atlantic and the Americas.

Other orbits

Many other orbits can be envisaged to meet special mission requirements. For a telecommunications satellite system intended to serve only the northern high-latitude regions, in which geostationary satellites are seen only at low elevations (10 deg. over the horizon at 65 deg. latitude), it can be advantageous to select a highly eccentric orbit with a period of about 12 hours and with an apogee (at about 39 000 km) high in the Northern hemisphere. Such an orbit has been selected by the Russians for their *Molniya* satellites. The Canadians, however, in meeting rather similar requirements, have chosen a geostationary orbit for their *Telsat* project.

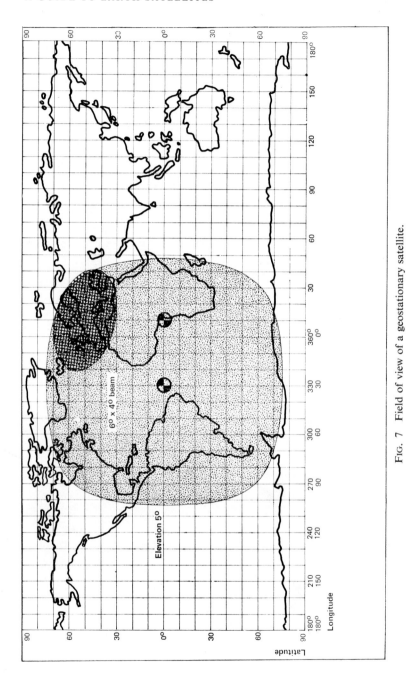

Fig. 7 Field of view of a geostationary satellite.

Satellite design

Satellites are almost always tailored to meet the mission requirements of specific applications, discussed in more detail in the chapters that follow. Their design is primarily influenced by the combination of
(1) the characteristics of the useful payload or equipment,
(2) the geometry of the selected orbit, and
(3) the characteristics and performance of the launch vehicle.
I shall confine my remarks to sub-systems all satellites have in common.

First, structure, which determines the shape of the satellite and provides for the installation of all the other sub-systems and of the payload. During ascent, it carries all loads of the satellite to the last stage attachment points (or ring). The structure must be light (15–25 per cent of total spacecraft weight), and must avoid resonances which would amplify by a factor exceeding 4 to 6 the vibrations communicated by the rocket.

Spacecraft structures fall into two broad categories. The one most frequently used comprises a rigid core, such as a stiff torsion tube carrying all loads transmitted by plateaux (to which equipment is attached), with outriggers to support the other structural elements such as solar panels. In the alternative, a box structure, most of the loads are carried by the outer surfaces. It is used when large pressurized volumes are required, for instance for manned flight.

Second, thermal control, the function of which is to keep within suitable limits the temperature of the other sub-systems. A satellite is heated by absorption of solar radiation (flux equal to 0.14 W/cm^2) and by internal power dissipation. Part of the solar flux is reflected by the outer skin of the satellite, which also radiates heat in proportion to the fourth power of its absolute temperature. Overall temperature control is achieved by selection of surfaces with appropriate absorption and reflection characteristics.

Third, power supply. In almost all Earth satellites, silicon solar cells are used to convert solar into electrical energy, with an efficiency of conversion of the order of 9 per cent at the beginning of life (dropping to about 8 per cent after one year in orbit and 7.5 per cent after five years under the influence of radiation and micro-meteorites). A square centimetre of cells installed perpendicular to the solar flux of 0.14 W/cm^2 produces an end-of-life power of about 0.0105 W. Solar cells today are relatively heavy, and must be mounted on rigid panels. But flexible supports under development promise large, lightweight surfaces capable of providing several kilowatts at a weight of about 60 W/kg.

Fourth, attitude control, the function of which is to maintain the satellite in the required attitude with respect to the Earth and the

Sun. It influences the overall design of satellites, which usually fall into one of two broad categories – spinning satellites or 3-axis stabilized satellites.

When the attitude of the satellite can be kept nearly constant over extended periods, it is often very convenient to spin it to obtain gyroscopic stability. This is especially the case when an axis must remain perpendicular to the orbital plane.

When the whole body of the satellite must be pointed in a direction varying with time, it is necessary to resort to 3-axis stabilization. When combined with a design which keeps the solar array oriented towards the Sun, significantly larger powers can be obtained compared with spinning satellites, at least for overall weights above 250 kg.

Fifth, attitude reference, for in order to stabilize the satellite in the required direction, we must know the attitude. This is the task of the attitude reference sub-system, which derives its information by measuring the angles between the direction of the Sun, the edges of the Earth, the brightest stars, and possibly some transmitters on the ground, and the spacecraft axis. Attitude can be obtained with an accuracy of the order of 0·1–0·4 deg. – good enough for most applications – using only solar sensors and infrared Earth sensors. Stellar sensors afford much higher accuracy but are more complex.

Sixth, station-keeping, for which a sub-system is required only when it is necessary to control closely the position of the satellite – as with the geostationary satellites. Propulsion, therefore, is needed to correct injection errors, so that the satellite can drift slowly in longitude for days, even weeks, then when the correct longitude position has been reached, be adjusted to a 24-hour orbit. Propulsion is also required to compensate for orbital perturbations, although these adjustments will need a tenth or less of the power of injection error demands.

Typically, geostationary satellites designed for a five-year life are provided with a propulsion capability of 150 to 300 m/sec, using small thrusters fuelled by monopropellants such as hydrogen peroxide or hydrazine. In the future, electric propulsion may be used in this role, saving the weight of perhaps 50 kg of hydrazine for a 500-kg satellite.

Communication sub-system

The seventh and final element is the communication sub-system, which usually consists of telemetry, data storage and telecommand equipment, together with an aerial system.

Telemetry: Information, gathered on-board (or, in the case of tele-communication missions, received from the ground for relay by the satellite) is transmitted to the ground via telemetry either in analogue or digital form. Even at modest power, substantial amounts of information can be transmitted to the ground. In the most frequently used VHF band it is possible, with a 5 dB margin and an omni-directional spacecraft aerial, to transmit to the ground:

120 000 bits/sec W radiated from a   500-km orbit.
60 000 bits/sec W radiated from a 1000-km orbit.
25 000 bits/sec W radiated from a 2000-km orbit.
600 bits/sec W radiated from a geostationary orbit.

One watt radiated corresponds to 2·5 watts consumed at VHF frequencies, but the efficiency declines at the higher frequencies needed for higher bit rates. For high bit rates, in excess of 10 000 bits/sec, it becomes difficult to find in VHF the necessary bandwidth (1 bit/sec requires approximately 1 Hz of bandwidth), and it is necessary to select a higher frequency available for space uses, such as the S-band around 2·2 GHz.

Data storage: In the simplest case, with the satellite in a low orbit, a significant fraction of the information would be lost because the satellite is only intermittently in view of a ground station. The answer is to record and store information on-board and transmit it, on command, when the satellite is in view of a ground station.

For relatively small amounts of information, reliable magnetic core memories can be used. Their power consumption can be rather high – 1 W at 200 bits/sec, 50 W at 200 000 bits/sec – but their use is mainly limited by weight for it is possible to store only up to 30 000 bits per kg. To store, say, a quarter million bits or more it is often necessary to use magnetic tape records. A typical tape-recorder would be capable of storing 2·8 million bits, have a weight of 3 kg, a power consumption of 0·85 W when recording and 1·50 W when playing back, and a speed ratio of about 30, permitting it for instance to play back in four minutes the information gathered during a complete 120-minute orbit. The main problem associated with tape-recorders is one of reliability – their probable lifetime does not exceed 6 months.

Magnetic tape-recorders are limited at present to a few million bits; but video recorders, for the storage of larger amounts of information, are under development.

Telecommand: Orders have to be transmitted to the satellite, for instance to command attitude manœuvres and recorder playback. For this purpose the spacecraft is always equipped with a command receiver and decoder. One widely-used command system, the NASA tone digital standard, operates in VHF at 148 MHz, providing up to 70 different commands per tone used.

Aerial system: At 136 MHz, the wavelength is 2·2 metres – that is comparable to the dimensions of the satellite – and it is difficult to obtain significant gains unless very special and bulky designs are used such as the deployable parabolic aerial, 10 m in diameter, selected by NASA for ATS-F and -G technological satellites. Often satellites transmitting in VHF are equipped with nearly omnidirectional aerials (−4 dB holes) of the crossdipoles or turnstile types, which serve simultaneously for telemetry transmission at 137 MHz and command reception at 148 MHz. When higher frequencies are used, either for transmission or for reception, the wavelength rapidly becomes much smaller than the satellite dimensions and it becomes, therefore, more and more difficult to achieve a nearly omnidirectional pattern (for instance, the wavelength at 2·2 GHz is 13·5 cm and holes of the order of −10 dB have to be tolerated).

The advantage of the parabolic reflector is the ease with which it can increase the amount of energy received.

The latest idea is the phased-array aerial, an all-electronic system consisting of a group of radiating elements, fed with the proper phases relative to one another to focus the radiated energy in the required direction. Phased-arrays are under development for use both on-board and on the ground.

Lastly, I must mention 'house-keeping', the sub-system which serves essentially to monitor and protect the other on-board sub-systems. It allows us to measure, via telemetry, temperatures at key locations, the values of current and voltages, the status of relays, the pressure in tanks, and the contents of certain registers. It also ensures protection, by means of relays, against short-circuits.

System reliability

System reliability is of utmost importance as it determines the number of launches required, and consequently the economics of an application mission. It is achieved through design reliability and quality assurance.

First, design reliability: A satellite consists of sub-systems which must all function properly, and can thus be considered to be installed in series, so that the overall satellite reliability R_s is the product of the individual reliabilities R_i of all sub-systems:

$$R_s = \prod_i R_i$$

with reliability R_i related as follows to the probability of failure F_i of sub-system i,

$$F_i = 1 - R_i$$

In turn, each sub-system i consists of a series of parts (resistors capacitors, transistors, diodes, integrated circuits) each with its own failure rate $\lambda_{i,j}$ (number of failures per billion operating hours), and the sub-system reliability R_i is related to the time of operation T and to the sum of the failure rates by the following relationship:

$$R_i = \exp\left(-T \sum \lambda_{i,j}\right)$$

Design reliability is mainly achieved by very careful circuit design, selection of space-qualified electronics and the derating of these parts.

There is, however, an upper reliability limit which is rather rapidly reached by applying only these standard precautions when higher values are required. So we must resort to redundancy of parts, even of whole sub-system. Indeed the overall reliability $R_{i,i}$ of a block of two elements installed in parallel is:

$$R_i = 1 - (1 - R_i)^2$$

R_i being the reliability of each separate element. Doubling up an element with a reliability of 0·90 yields an equivalent reliability of 0·99.

Nowadays, it is possible in many cases to reach fairly high overall reliabilities, of the order of 0·60 to 0·80 after 5 to 8 years – at least when no delicate instruments such as magnetic tape recorders have to be used. Nor, in general, is redundancy very costly in weight – it seldom requires more than 10–20 per cent of the overall satellite weight.

Second, *quality assurance:* In addition to design reliability, overall reliability also requires a very strict programme of quality assurance that aims, for example, to ensure that manufacture and integration take place under strictly controlled conditions of cleanliness; all processes are certified and controlled; mechanical electronic parts used are space qualified; all failures are systematically reported and analysed; design reviews are run at critical times of the project.

Finally, *system reliability:* Even if all these precautions are taken, failures will nevertheless occur. Their statistical prediction is of great importance for forecasting the system economics.

Planning a satellite system

The complete development cycle of an Earth satellite, from the early identification of the mission to the launch, can take four to eight years or more. The first part of this cycle, seldom less than one year and often much more, is devoted to the preparation of the project, including:
- Precise definition of the mission in collaboration with the users.
- Detailed feasibility studies.
- Assembly of a project team.

● Go-ahead decision by the space agency and by the users' agencies, and the consecutive integration of the project in the budget.

● Preparation of the project specifications, for ground and communication equipment as well as the space segment.

How long this preparation phase continues will be influenced above all by interest in the mission, by the readiness of the users' agencies to utilize a new space system, and by the economic situation and probably political problems – especially when the implications are international.

It is difficult to provide a general description of this phase, which differs considerably from project to project – except inasmuch as all five activities listed above must be complete at its end. It is also preferable to see, as far as possible, that the system is within the state of the art. If not, it is most advisable to initiate, during the preparation phase, the development of the long lead-time items; this is often done with research and development funds, rather than project funds.

The next phase is the hardware development phase, influenced chiefly by the contractual approach used in the simplest case a single group of firms, under the leadership of a prime contractor, is selected through a tendering operation which takes about 9 months from the date of issue of the requests for proposals (3 months for the preparation, by industry, of the proposals; 3 months for their evaluation; 3 months for selection and contract negotiations).

Although the quickest approach it is nowadays used only for the simplest projects for it does not permit full advantage to be taken of competition, both technically and in costs.* More frequently, the hardware development phase is split into two or three parts – known as phased planning – and several groups of contractors are selected to perform competitively the first part or parts. The best group is then given the final hardware procurement contract. Typically, one would spend 10–20 per cent of the satellite costs with two groups of firms before selecting the winner.

In general, it is fair to say that increasing competition is likely to improve the technical characteristics of the system, and to reduce the risk of slippages and cost overruns. But inevitably it lengthens the planning by at least 3–6 months, and increases the overall costs by at least about 5–10 per cent.

The hardware development phase almost always includes:

● Detailed design of the spacecraft (4–8 months).

● Manufacture and test of a mechanical and a thermal model (3–6 months).

* NB. It is not often possible to place a fixed-price contract for satellite development.

- Breadboarding of the subsystems (2–6 months).
- Manufacture and test of engineering models of all subsystems (3–6 months).
- Integration and test of an electrical model, using the engineering models (2–6 months).
- Manufacture, integration and qualification testing of the prototype (8–16 months).
- Manufacture, integration and testing of the flight unit (6–12 months).

Often, too, the list includes manufacture and possible integration and test of a spare flight unit (sometimes the prototype is used as the spare, despite the fact that it has been overtested).

In principle, it would be desirable to plan in series the engineering models, the prototype and the flight unit. But this would almost always result in excessive delays and costs. It is general practice to accept some overlap between the activities and the consequent risks of having to introduce modifications when testing during one subphase identifies a weakness while the hardware of the following subphase is already finished.

A very important planning restraint is that delivery of high-reliability electronics, desirable for the prototype and mandatory for the flight units, can take up to 12 months.

The hardware development phase normally takes from 2 to 4 years, at the end of which the satellite is ready to leave for the range, 2 to 8 weeks before the launch.

How much?

To a first approximation, the development and manufacturing costs in industry of Earth satellites are proportional to their weight, with the coefficient of proportionality lying between $100,000 and $160,000 per kg, depending upon various factors such as:

- Past experience of contractors – often entire sub-systems already developed can readily be adapted.
- Design complexity – spinning satellites with body-mounted solar cells can be expected to be near the lower value, and 3-axis stabilized satellites near the higher.
- Payload complexity – costs increase rapidly with the complexity of payload instrumentation, even when this does not influence the general design of the satellite.
- Project duration – rapid developments are likely to cost significantly less than long ones but they are also associated with higher risks of delays and cost overruns.
- Overall weight – a second-order tendency exists, nevertheless, towards a decrease in specific costs per kg with increasing space-

21

craft weight, because several sub-systems are little affected by satellite size.

By way of example, a 270-kg spinning satellite with a despin aerial would probably have a specific cost of $120 000 per kg, corresponding to a total contract cost of $33 million. This total for the main satellite contract would include, first a competitive project definition phase run by two groups of firms. Second, it would include the development phase, including manufacturing and testing of breadboards and engineering models of the various sub-systems; of mechanical, thermal and electrical models of the entire satellite; and the qualification prototype. Third, it would cover manufacture and acceptance tests of one flight unit and one spare flight unit (one additional flight unit would cost about another 15 per cent of the industrial costs). And last, it would cover participation of the contractors in the launch operations.

But the total cost I have given would not include, in many instances, development and qualification of the payload instrumentation such as the telecommunication repeaters and tubes or an infrared scanning radiometer. It would not cover the internal costs of the space organizations, such as ESRO, NASA or INTELSAT, which monitor the project and often provide the large test facilities such as the solar simulators; and after the launch, control of the satellite in orbit, through a tracking, telemetry and tele-command network. Such costs often amount to 30 to 50 per cent of the satellite contracts costs. Nor would it cover the costs of the apogee motor, when needed (although this item is relatively cheap when no extensive new developments are required). Finally, it does not include the launch costs – the rocket and the use of the range facilities. A Delta costs about $6·5 million, a Centaur about $14 million. All these additional costs, can equal and even exceed the cost of the main satellite contract.

Even then, the bill of some $70 million from our example, still does not include the investment and running costs for the ground facilities (telecommunications station; air traffic control centres, meteorological centres), necessary for a specific application.

When all these costs are added up, including those required to maintain the system in orbit, the average overall costs for developing, launching, operating and maintaining a single application satellite in orbit over a period of 5–10 years generally exceed $10 million a year, and often reaches several times this figure.

CHAPTER **2**

Communication Satellites

J. K. S. JOWETT
Post Office Telecommunications Headquarters

The 1960s saw the development of a completely new communications medium. What, at the beginning of the decade, was no more than a concept in the minds of a few engineers had, by the end, become a fully commercial system providing global communications. Moreover, this global system was of high capacity, using satellites positioned in the geostationary orbit to interconnect many highly specialized Earth stations operating in more than thirty countries of the world. Other systems also, some for military communications, were planned or in use by 1970 and the prospects for future large-scale development are extremely bright.

The first really significant step in this process was taken with the successful launch of *Telstar*, the experimental satellite designed by the Bell Telephone Laboratories and placed in an elliptical inclined medium altitude orbit by NASA in 1962. *Telstar*, and the succeeding NASA satellite *Relay*, first clearly demonstrated the technical potential for direct, wide-band communications across the oceans of the world. One of the many wise decisions made in preparing for the *Telstar* tests was the choice of a frequency band around 6 GHz for the Earth-to-satellite path, and of a frequency band around 4 GHz for the satellite-to-Earth path. These bands – 5425-6425 MHz and 3700-4200 MHz – have become the basic frequency bands for *Intelsat* systems and have proved to be of the greatest value for this type of telecommunications system.

But these early satellites, singly, could provide at most a 45-minute period of simultaneous visibility from both sides of the Atlantic during each orbital period of about five hours. It was only with the experimental operation of the *Syncom* series of satellites in 1963 that the practical possibilities of building a system using a few satellites in the so-called geostationary orbit, of much higher altitude, began to be considered closely. The orbital period of a satellite is a function of altitude, as shown in Fig. 8, and becomes equal to the Earth's sidereal period of rotation (about 23 hours 56 minutes) at an altitude of 35 786 km. If launched in an orbit rotating from west to east in the Earth's equatorial plane, a satellite at this altitude will keep in step with the Earth's rotational movement, and will appear

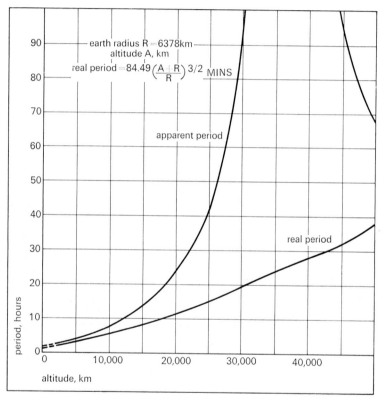

FIG. 8 Orbital period of a satellite in circular equatorial Earth orbit.

to be stationary to an observer on the Earth's surface. Certain perturbations of satellite motion occur due to solar and lunar gravitational attraction and to the non-uniformity of the Earth's gravitational field, but the *Syncom* tests demonstrated that these could be compensated for by the controlled ejection of gas from the spacecraft.

This left, basically, only one unresolved question: the effect that a high-altitude satellite would have on normal two-way telephone conversations, for which the one-way transmission time would be around

270 milliseconds. *Early Bird,* launched in 1964 (later renamed *Intelsat* 1) showed that our fears were unfounded.

Since that date, all *Intelsat* satellites have employed the geostationary orbit with a high degree of success. Much of the discussion in this chapter relates to the principles and operation of this type of

FIG. 9   *Intelsat* 3, 150-kg satellite on which the first global satellite system is based.

satellite within the *Intelsat* global network, of which a fuller description is given later. Here it is sufficient to say that the first global system, employing *Intelsat* 3 satellites over each of the three main oceans, was established in 1969. Each of these satellites (Fig. 9) can provide some 1000 simultaneous telephone circuits and a common television channel between numerous Earth stations, each having aerials 25–30 m in diameter.

25

Other systems of communication satellites have been built in recent years, including a Russian system, known as *Orbita*, using a number of *Molnya* satellites operating in a highly elliptical 12-hour orbit, inclined at about 65° to the Earth's equatorial plane. This system is primarily used for the distribution of television programmes over a limited coverage area. The choice of this orbit arose from the need to provide long east–west connections at high northerly latitudes. However, the USSR is also planning the use of geostationary satellites.

Geostationary satellite coverage

While the potential coverage of the Earth's surface from a single geostationary satellite is extensive, it cannot include regions near the Poles and, even for low-latitude regions, at least three such satellites are needed to give world-wide coverage. In practice, for a world-wide system (excluding very high-latitude regions where traffic requirements are, fortunately, negligible) rather more than three satellites may be desirable in order to provide mutual visibility via a single satellite between all communities having strong commercial relations with one another.

Fig. 10 provides curves showing the limits of coverage possible from a geostationary satellite whose location is precisely controlled, for various assumed minimum angles of elevation of the satellite as seen from the Earth stations. A working elevation angle limit which has so far been adopted is 5 degrees, but future systems employing much higher frequencies of operation may need to limit at 10 or even 20 degrees to avoid problems of excessive attenuation of signals passing at low angles through heavy rain or dense cloud. Typical coverage areas for the *Intelsat* systems are given later in this Chapter. Atmospheric refraction can theoretically extend the coverage but this is significant only for very low angles of elevation of the order of 1 or 2 degrees. Refraction therefore plays no significant part in extending the actual maximum coverage of practical systems. If we take account of the fact that a typical Earth station aerial of 27 m diameter might have a gain of about 58 dB, the true signal attenuation, under normally prevailing free-space loss conditions, ranges from 125·5 dB to 126·8 dB.

Although this is a very high attenuation which called, initially, for extremely low-noise receiving systems and the use of wide-band modulation techniques, there is the great advantage that the path attenuation to the various Earth stations differs little within the zone covered. Furthermore, signal fading – so destructive of short-wave communications via the ionosphere – presents no serious problems in the microwave bands so far used.

The consequence is that a geostationary satellite system operating in the 4 and 6 GHz bands can be designed with confidence to meet the known radio-path conditions. Moreover, virtually all Earth stations within the system can be designed to a standard performance specification – a substantial advantage when establishing a system for international communications. Stations working at or near the

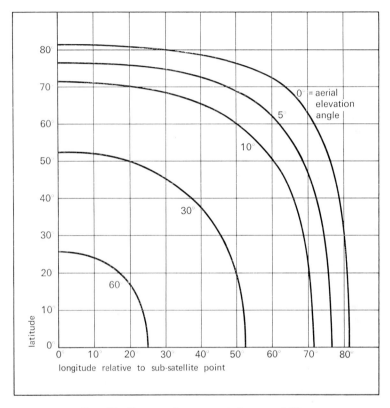

FIG. 10   Coverage from a geostationary satellite.

fringe of the service area may, it is true, need to be supplied with rather more satellite power than those operating at higher elevation angles. But this is a minor problem which may one day be overcome when satellite aerials can be designed with specially shaped beams to give a more or less constant power flux density at all parts of the illuminated surface of the Earth.

27

Frequencies and modulation methods

These considerations lead to some fundamental choices concerning the systems of transmission in both directions between an Earth station and a communication satellite. Because of the low transmitter powers so far it is the 'down' path from satellite to Earth station which presents the greater design problem. It calls for a method of signal modulation in which bandwidth is exchanged for power. Theoretically it would be possible for a modulation method to be used on the 'up' path which made more economical use of bandwidth but, with few exceptions, systems have used the same signal processing on the up path as on the down path. This simplifies the design of the satellite, which then needs only to provide signal amplification (of the order of 100 dB), apart from a straightforward change of radio frequency within the satellite to provide sufficient separation between the incoming and outgoing frequencies.

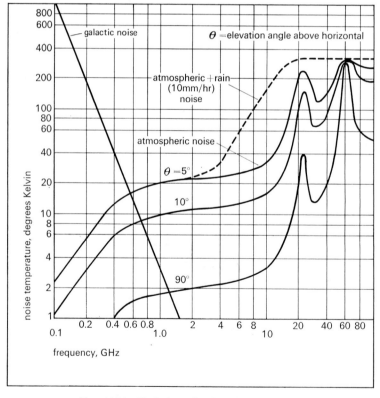

Fig. 11(a) Variation of noise with frequency.

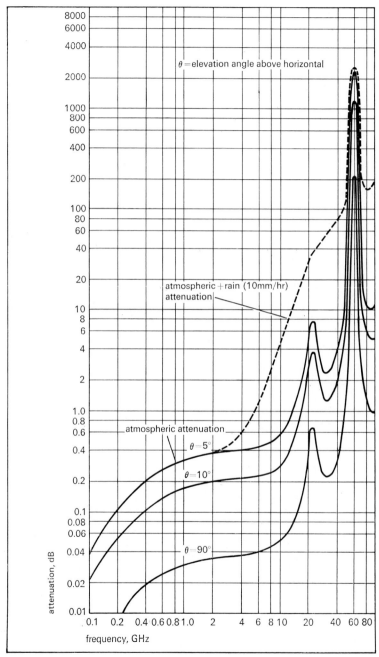

FIG. 11(b) Attenuation of signal with frequency.

The frequencies to be used for the down path in particular must be in a region of the frequency spectrum in which naturally-occurring noise is very low and the possibilities of substantial atmospheric attenuation of the signal are remote. Curves of Fig. 11(a) show the variation of galactic and atmospheric noise with frequency for a clear atmosphere. Under conditions of precipitation the noise levels at the higher frequencies could rise substantially. Curves of Fig. 11(b) illustrate the signal attenuation, over and above that due to free-space spreading of the wave, which takes place under clear-sky conditions. Again very considerable increases of attenuation can take place, especially above 10 GHz, in heavy rain.

For these reasons frequencies between about 1 and 10 GHz have thus far been preferred for operation with communication satellites. *Intelsat* has made extensive use of the 4- and 6-GHz bands; and it is expected that future Russian satellite systems will use similar frequency bands, although the initial *Molnya* satellite operated at much lower frequencies. A Space Frequency Conference organized by the International Telecommunication Union in 1963 allocated a number of bands between 3·4 and 8·7 GHz for communication satellite systems, these bands being those already partially occupied by terrestrial wide-band radio-relay systems. Since all bands below 10·5 GHz had been allocated to the various terrestrial radio services at a previous conference, considerable sharing of the frequency spectrum by the new space services was unavoidable. It was judged, rightly, that the most efficient arrangement would be that in which communication satellite systems shared bands with radio-relay links, which use highly-directive, low-power transmissions generally in a horizontal direction. To regulate the shared use of the allocated frequency bands, certain technical criteria have been internationally adopted, the main ones being a limitation of the equivalent isotropically radiated power of radio-relay stations to 55 dBW, and a limitation of the power flux density which may be set up by a horizontally arriving satellite signal at the surface of the Earth to -152 dBW/m^2 in any 4-kHz band. Somewhat higher power flux densities may be set up at the higher angles of elevation.

For the *Intelsat* 3 system it turns out that a total of some 160 telephone channels could be transmitted on a 1-watt frequency modulated carrier. In practice, an operating point a few decibels above the threshold value is desirable, and at present this margin is being achieved by the use of threshold-extension FM receivers at the Earth station. But the advent of more powerful satellites will make this provision less necessary.

Signals may alternatively be assembled in digital form, such as pulse code modulation (PCM), and progress in developing such types of system is expected to be rapid. On the other hand, single side-band amplitude modulation, though operationally attractive, cannot easily provide the desired performance and its use in systems giving a high standard of performance is unlikely.

Multiple access

An essential and very advantageous feature of modern communication satellite systems is the facility to provide simultaneous access to one satellite from a large number of Earth stations within a common coverage zone.

The provision of full multiple access with high capacity had to await the launch, over the Atlantic, of the first *Intelsat* 3 satellite in 1968. With the increased satellite power and aerial gain, transmission problems are easier than in the case of *Intelsat* 1, and carrier deviations are therefore lower. Typical operating conditions are shown in Table 1.

TABLE 1. Typical satellite operating conditions

Number of channels per carrier	R.F. bandwidth	Satellite output power
24 (telephony)	5 MHz	0·3 watts
60 (telephony)	10 MHz	0·6 watts
132 (telephony)	20 MHz	1·2 watts
1 (television)	40 MHz	4·0 watts

Preparation of a frequency plan to allow many carriers to operate through *Intelsat* 3's two wideband transponders without excessive intermodulation presented a complex problem. However, by under-driving the final TWT amplifiers and by requiring Earth stations to add an artificial low frequency 'dispersion' signal to prevent carrier energy from being concentrated in any small part of the spectrum, intermodulation noise is kept to within acceptable limits. Nevertheless this type of noise has to be allowed for in an overall noise budget (see Table 2).

This total noise value of 10 000 pW corresponds to the recommended CCIR and CCITT performance for a single-hop satellite circuit, and is assumed to be measured at a nominal zero level (that is 1 mW) point in a telephony circuit. The corresponding signal-to-noise level is thus 50 dB but since normal speech levels fall below the standard line-up level the true speech signal-to-noise ratio is generally less.

As the development of higher power satellites takes place, and the

TABLE 2   Typical noise budget for a telephone circuit via *Intelsat* 3

Up-path thermal noise	1400 pW
Satellite transponder intermodulation noise	2400 pW
Down-path thermal noise	4200 pW
Multiplex equipment noise	1000 pW
Noise due to possible interference from	
radio-relay links	1000 pW
TOTAL	**10 000 pW**

need lessens to use very large bandwidths in order to meet performance requirements, the effects of intermodulation distortion – not
only in the satellite but also in earth station transmitters – will become more pronounced, as in the *Intelsat* 4 system. The outcome
may be a swing away from analogue methods of modulation to
digital systems, including those using time division, and to the
design of more complex satellites.

Earth station design

The design of an Earth station to operate in a satellite communications system is, of course, closely related to and dependent upon the
design of the satellite with which it is to operate and the type and
quantity of traffic which it is to carry. Of first importance is the
question of aerial aperture size and this is governed by several
factors, including:
(1) The power output of the satellite and, in particular, the power flux
density set up at the Earth's surface.
(2) The frequency bands used.
(3) The minimum safe aerial beamwidth, allowing for difficulties of
tracking, particularly under high wind conditions.
(4) Overall system economics, including both the Earth station cost
(which obviously increases with increase in aerial size) and the
charges for use of the space segment (which may decrease sharply
with increase in aerial size).

For certain types of system, for example some military systems and
future regional systems, aerial sizes of the order of 10 m or less
diameter may prove optimum. However, for the commercial global
system of *Intelsat* the state of satellite technology and the overall
system economics have required the use of larger aerials, 25 to 30 m.
In the 4- and 6-GHz bands in which they have been used such
aerials provide 3-dB beamwidths of the order of 0·2 degrees. Experience has shown that tracking problems can be solved even with
these narrow beamwidths provided there is some form of mechanical

or electronic autotrack arrangement, to compensate for movement due to wind gusts.

With this size of aerial working to *Intelsat* 3, transmit carrier powers ranging from 15 to 200 watts are adequate to provide for signals ranging from 24 channels of telephony to CCIR standard colour television signals. Some stations, such as that at Goonhilly, employ 5 kW TWTs having a 500 MHz bandwidth and capable of handling several telephony carriers and a television carrier. Other stations operate with a number of lower-powered klystrons, each covering the much smaller bandwidth necessary for each individual carrier transmitted.

On the receiver side, conditions are much more critical due to the relatively low powers transmitted from the satellite. The masers used in the early *Telstar* experiments have given way universally to cooled parametric amplifiers.

There are a number of components of noise to take into account and a typical breakdown is given in Table 3.

TABLE 3.   Typical noise components in the receiver

Clear sky temperature (at 10° elevation)	15 K
Waveguide attenuation noise	20 K
Spillover noise (at 10° elevation)	10 K
Cooled parametric amplifier	20 K
TOTAL	65 K

Excess noise occurs, of course, whenever rain causes atmospheric attenuation, but this effect has proved to be small in the 4-GHz frequency band. At frequencies above 10 GHz, attenuation may rise considerably on occasions and this would be associated with a substantial increase of sky noise temperature.

This noise breakdown is typical of an aerial system using the Cassegrain arrangement in which the paraboloidal main reflector is illuminated via a hyperboloidal sub-reflector. Such arrangements have now been widely adopted and allow the main feed horn illuminating the sub-reflector to be mounted through a hole at the centre of the main reflector, and the low noise receiver to be mounted very close to the feed horn. In order to meet the stringent gain requirements the surface profile of the main reflector must be maintained to within about 1 mm of the theoretical profile.

In the period of experimental communication satellites aerials were built so that they could follow moving satellites in any part of the sky. Even with nominally stationary satellites it is still necessary to have a capability of steering the aerial to different points in geostationary orbit and to follow satellites whose orbits depart slightly ᐟ from the ideal. In any case, operational requirements may, during the lifetime of an Earth station, occasionally require an aerial to transfer from one satellite to another in a quite different orbital location. Full steerability requires that the aerial has two perpendicular axes of rotation and, almost without exception, the chosen arrangement is the azimuth-elevation mount. Fig. 12 shows the Marconi-built aerial

FIG. 12   Bahrain Earth station built by Marconi.

of the Bahrain Earth station, the main features of which are fairly typical of a number of present and planned Earth stations. Here the azimuth bearing is housed inside the top of a circular concrete tower some 18 m high and 5 m diameter at the top. The azimuth drive is through a fixed bull-ring gear on the top of the tower. The elevation drive and bearings are mounted above the azimuth bearing.

Also carried on this part of the aerial is the low-noise receiver which may or may not move in elevation. The high-power transmitters and other radio equipment are housed in the fixed tower and connected to the feed horn by a flexible waveguide. The motors driving the aerial in azimuth and elevation are part of a servo control system for which the error signal is dependent on any difference between the direction of the electrical axis of the aerial and the direction of the arrival of signals from the satellite.

A wideband link, capable of simultaneously carrying both hundreds of telephone circuits and television signals, connects the Earth station to a central national terminal. Such wideband systems normally comprise multiple-hop microwave radio-relay links; for example, a 13-hop link has been provided between Goonhilly Down and London, where connections are made to the national telephone network and to the Post Office network feeding BBC and ITA television broadcasting stations.

The cost of the special interface equipment at an Earth station and of the communicating link to the national terminal is not a small item and from the economic point of view an Earth station should desirably be sited close to the national terminal. In densely populated and well-developed countries such as the United Kingdom, a number of factors militate against this, in particular the extensive use already made of the microwave bands for the terrestrial radio-relay network. Together with the need to have an Earth station site well clear of all sources of radio-interference, it led the Post Office to choose a site for their commercial station in Cornwall. In less intensively-developed countries, however, much shorter distances suffice, particularly where the Earth station can be shielded by a natural 'bowl' of surrounding hills.

Communication satellite design

All spacecraft, because of their high constructional and launching costs, are designed to have a high degree of reliability. Few, however, have the special requirement of commercial communication satellites that they must perform uniformly well for 24 hours of every day over a period of several years. Moreover, this performance must be achieved within a budget imposed by commercial constraints such as competition from submarine cables. An indication of the costs of satellite production and launching is given in Fig. 13.

An operational lifetime of at least five years is a further require-ment which enters into the design of all parts of the spacecraft. For example, solar cells, universally adopted for the primary power supply, must still be delivering adequate power at the end of the design life-time. The craft also needs batteries, because a geo-stationary satellite is exposed to solar eclipses for brief periods. The design of batteries to provide the necessary long life-time is in itself

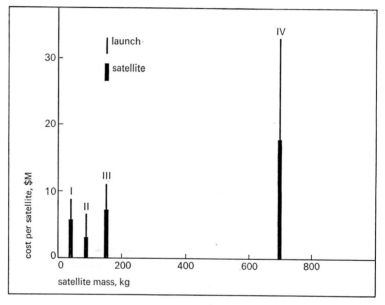

FIG. 13   Costs of building and launching *Intelsat* satellites.

a major problem. In addition, enough fuel must be carried for satellite positioning and orientation to allow for all foreseeable satellite manœuvres within its lifetime, together with some margin for contingencies.

Duplication, or dual redundancy, is highly desirable for all essen-tial sub-systems including the communications package itself. Nor-mal exceptions are the communication aerials, which should be sufficiently robust, and the mechanical bearings between the spinning and de-spun parts of some current satellite designs. These bearings, though open to space and therefore not easily lubricated, cannot readily be duplicated and therefore present one of the undoubted hazards in the design of this type of spacecraft. Full duplication of all parts of the communications package may not be essential in, for example, designs employing multiple narrow-band transponders operating on different frequencies. The premature loss of one or two out of twelve such transponders would clearly not be catastrophic although lessening the usable capacity at end of life. On the other hand, some crucial parts of the communications package, such as the broad-band receiver front-end which may be common to all communication signals, or the power output stages which generally employ TWT amplifiers, need in some cases to be triplicated in order to provide sufficient reliability.

A considerable amount of information is accumulating on the reliability of specific components and this is used to derive circuit and sub-system networks which will provide the necessary overall redundancy and reliability. With the exception of TWTs which

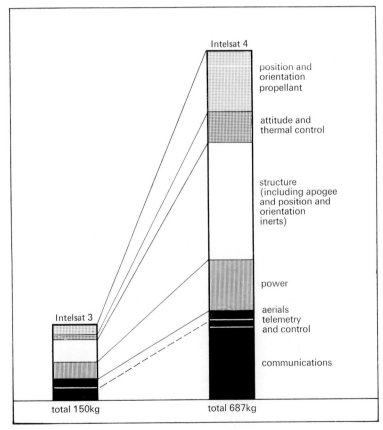

Fig. 14   Comparison of sub-system masses in *Intelsats* 3 and 4.

hitherto have generally been necessary to produce the r.f. outputs, solid-state components have been used with a growing emphasis on micro-miniaturization to reduce weight. Exceptional precautions are taken in the factory to ensure cleanliness of production techniques and to organize special selection and quality control procedures. The testing of the completed communication satellite is also a very rigorous and costly exercise carried out both within and without a simulated space environment and both before and after vibration tests simulating the effects of the several firings of the launcher rockets.

A comparison of the breakdown by mass of the more important sub-system of the 150-kg *Intelsat* 3 and the 690-kg *Intelsat* 4 spacecraft is shown in Fig. 14. This illustrates the still fairly small part which is taken up by the communications package and aerials which are the directly usable sub-systems of this type of spacecraft.

A view of the *Intelsat* 4 spacecraft is shown in Fig. 15. Like its

FIG. 15 *Intelsat* 4, 687-lb satellite, scheduled for service in 1971. (*Courtesy of Hughes Aircraft Company.*)

predecessors, this satellite is designed to move in a 'wheel' motion when in the 24-hour orbit. The curved surface of the 3-m diameter spinning drum is covered with some 45000 silicon solar-cells and many of the spacecraft sub-systems are mounted within the drum and rotate with it. The aerials and the twelve transponders are mounted on a platform which is driven by motors so as to precisely de-spin the aerials and hold them at all times in the necessary earth-oriented direction. Synchronization is provided by means of Earth sensors mounted on the curved surface of the spinning drum, these provide an electrical impulse every time the edge of the Earth comes into or goes out of view.

Some of the transponders are connected to the large dish transmitting aerials which provide so-called 'spot' beams, about 5 degrees wide. When used over the Atlantic one of these beams covers Europe and the other the Eastern seaboard of North America and thus provide a high power flux density to Earth stations in these areas at the cost of only a modest transmitter power. The two small horn aerials have a 17·5 degree beamwidth and provide for communication between all Earth stations in the total area of visibility.

The basic reason for spinning the spacecraft is to provide, in a relatively simple way, stability of spacecraft attitude against possible roll and yaw motions. This is easier to achieve when the spacecraft is shaped more like a disc (as in the case of the small *Intelsat* 1) than when it is roller-shaped as it is with the larger spacecraft. Other design approaches employ a non-spinning 'body', which can be controlled in attitude by automatically energized gas jets or, partially at least, by a heavy but small spinning wheel the bearings for which are protected from the difficult space environment. This type of spacecraft design may have the solar cells mounted on its body, or on fixed or rotatable vanes or panels which protrude from the main structure of the spacecraft. Undoubtedly, new approaches to design will be developed in the next decade. The experimental NASA ATS-F and ATS-G satellites will use a 3-axis stabilized approach. Another approach is to employ long booms for gravity-gradient stabilization but this technique has hitherto been tested successfully only in low- and medium-altitude satellites.

Pending the development of more complex communications packages, possibly involving de-modulation and re-modulation of signals in the satellite, a simpler approach involving multi-carrier transmission through wide-band transponders has been generally followed. The use of wide-band frequency modulation and of transponders operating in a quasi-linear mode has met the requirements without excessive intermodulation distortion between carriers. The current tendency is to swing away from the use of one or two very wide-band transponders and to employ a larger number of narrower-band transponders, connected as desired to aerials in response to operational requirements – which may, of course, change during satellite lifetime. These developments in turn call for the development of highly reliable microwave switches to set up the various operational conditions by command from the ground. A major problem in this approach to the communications package design is the development of a frequency plan satisfactory to all of the Earth stations involved. Inevitably, frequent changes take place in Earth station traffic requirements and therefore in the frequency plan. Such operational requirements emphasize the need for satellites with a greater degree of flexibility, not only as to the use of transponders but also in the pointing of 'spot-beam' aerials and in the methods of signal modulation which may be used.

Typically, communication-satellite receivers work with an input system noise temperature of around 2000 K, though this may fall to 1000 K with time. Output powers of the order of 6 to 10 watts are achieved by the use of TWTs, which will no doubt in time be replaced by solid-state devices. Analogue methods of modulation are beginning to give way to digital methods of signal processing but it remains to be seen whether this slow transformation will greatly change the form of the communications package on board the satellite. There are some who think that the communications satellite of the future will possess extensive signal-processing and switching facilities and may thus work almost as a telephone exchange in space.

Growth of a global network

The only existing global telecommunications network involving satellites is operated by *Intelsat*, a consortium formed by 11 countries in 1964 and now involving some 80 countries. Its first global system was established in 1969 with the satellites and coverage areas shown in Fig. 16. The satellites are owned and operated as part of the Space Segment by *Intelsat*, investment being shared in various proportions by each of the member countries. An annual charge is made for each half-telephone circuit operated via the satellite by any of the Earth stations within the system and these charges, plus charges for television-relay transmissions, etc., provide the revenue for the consortium. When management costs have been met, the revenue is shared between the member countries in proportion to their investment contributions.

Fig. 17 shows various aspects of the successful growth of this system since 1964, this includes the initial use of the smaller capacity *Intelsat* 1 and 2 satellites. The curves have been extended to show the expected developments up to 1975.

Although as can be seen from Fig. 16 a single geostationary

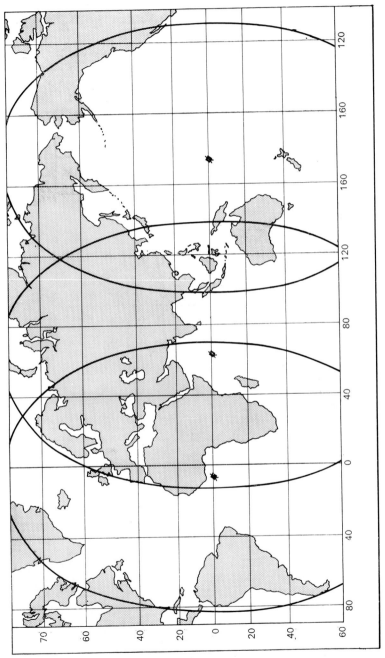

FIG. 16 Three operational satellite positions and coverage by *Intelsat*.

satellite – given full visible Earth coverage aerials – covers a large portion of the Earth's surface, there are inevitably difficulties in selecting the position in the orbit which will give optimum operational coverage in any particular situation. A classic example was the fixing

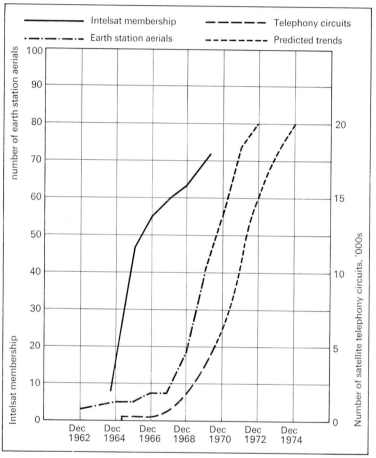

FIG. 17 Growth curves and forecasts for the *Intelsat* system.

of the Indian Ocean satellite location so as to allow intercommunication between stations as far west as Goonhilly Down and as far east as Japan and Eastern Australia. A location of around 62 degrees E longitude allows stations at Goonhilly, Yamaguchi (Western Japan) and Ceduna (near Adelaide) all to see the satellite at an angle of elevation exceeding 5 degrees.

A consequential requirement, however, is that the satellite position be held to within ± 0.5 degree both in latitude and longitude since otherwise elevation angles at these three stations would at times become too low and might degrade circuit performance. For such reasons it is to be expected that communication satellites of the future will need to be closely limited in their orbital position. This contrasts with the *Intelsat* 1 situation over the Atlantic, since it was permissible to allow this satellite to drift by ± 15 degrees or more from its nominal position before exercising one of its control jets to reverse its drift motion.

Control of longitudinal position involves small amounts of fuel but both the amount and the frequency at which control needs to be exercised depend upon the longitude concerned, being a minimum at longitudes of about 115 degrees W and 65 degrees E. Control of the orbital plane of a satellite is a different matter. This plane at the present epoch shifts about 0.9 degree every year due to the effects of the solar and lunar gravitational fields and a considerable amount of fuel is needed to maintain it close to the truly equatorial plane. One technique which has been successfully used is to arrange the launch sequence so that the satellite initially goes into a plane inclined at 1 or 2 degrees to the equatorial plane in such a sense that the equatorial plane is naturally reached after one or two years in orbit. By this means the need to use fuel for orbit plane correction may be avoided during the early years of operation.

The future use of satellites for long-distance communications seems assured though the form of international cooperation may be expected to evolve. Indeed, other systems are likely to develop but it is likely that these will be confined to the domestic needs of large countries such as the US and Canada and the middle-distance requirements of certain groups of countries or regions. Examples of the latter are the already existing *Orbita* system used by the USSR and some Eastern European countries and the possible development of a European system for television distribution and telephony.

So far as telephony, at least, is concerned the need to employ satellites for communication between neighbouring countries is very dependent upon the cost-competitiveness of satellites with other means of communication. Whereas the cost of terrestrial means of communication (whether via cables or microwave radio-relay links) increases with the total distances to be spanned, the cost of satellite communications is – within certain definite limits – little affected by distance. Hence there is, for any particular set of economic circumstances and traffic requirements, a limiting 'cross-over' distance beyond which satellite communications would be the more economic.

Future developments

Thus far we have concentrated attention largely on the type of communication satellite system which can provide long-distance telecommunications between virtually all parts of the world and, in the form of the *Intelsat* network, is already in commercial use. The consortium is, in fact, not only a main supplier of long-distance international telephone circuits but also provides the commercial means for relaying television programmes between many countries of the world. Television has been carried on the system primarily to relay programmes for ultimate broadcasting to the general public, but a demand may well grow for closed-circuit television over long distances; this also can be met by communication satellites. Furthermore, data signals are already being transmitted via satellites and the satellite medium is, in due course, likely to prove a very effective one for the transmission over wide coverage areas of wide-band as well as narrow-band data.

A particular feature of satellite communication is its ability to reach directly into any part of the world where traffic demand is sufficient to warrant the cost of building an Earth station. The extent to which the medium may be used is thus primarily an economic one but, in the ultimate, other factors such as the amount of usable frequency spectrum and of orbital space may exercise control.

Regional and national systems

Foremost among newer forms of systems are those intended to provide telecommunications over a restricted part of the Earth's surface only, such as national or domestic systems, or systems to cover multi-national areas such as Europe. These might use Earth stations having aerials of some 5 to 15 m diameter. Other future systems might provide, for example, for the distribution of television to a large number of specialized Earth stations – having aerials of perhaps 3–5 m diameter – within a single large national territory or group of territories. From such stations television programmes could be provided by means of land line connections to the homes of the general public. In the longer-term future a demand might arise for data to be provided via such a satellite distribution system. However, in both these examples, it must be remembered that more conventional terrestrial means could be used to distribute the signals and these latter means are likely to be more economic in smaller countries or areas.

Television broadcasting systems

Ultimately the possibility of television broadcasting direct from satellites to homes may be envisaged and various schemes have been proposed. The cost to the viewer must be small, and this requirement could best be met if satellite transmissions employed vestigial side-band amplitude modulation at frequencies in or near the existing UHF broadcasting bands. Such a system, however, would involve extremely high powers on board the satellite, which would in consequence be very costly both to build and to launch.

We now accept that some other modulation method, such as frequency modulation, would have to be used in order to conserve power and cost in the space segment. But this would entail more expense for each viewer, who would need to have a new i.f. and demodulator section added to his receiver as well as a special high-gain aerial and low-noise front end. Because of the relatively high costs of receiving such satellite transmissions – possibly of the order of £50 to £100 per viewer initially – it is unlikely that this new medium will be an attractive proposition in countries having extensive coverage already by means of terrestrial broadcasting facilities. Larger countries, in which television broadcasting is not yet highly developed, are more likely to find applications for satellite broadcasting and experimental satellite transmissions are already being planned for these purposes.

CHAPTER **3**

Meteorological Satellites

Dr B. J. Mason, FRS
Meteorological Office, Bracknell, England.

Satellites are already playing an important role in observing the weather on a global scale. During the last ten years the US has launched 20 operational polar-orbiting meteorological satellites in the *Tiros* and *Essa* series, together with four large *Nimbus* research vehicles and two geostationary applications technology satellites (ATS) that are largely devoted to meteorology. Having obtained cloud pictures and radiometric data on an experimental basis from *Cosmos* satellites since 1966, the USSR has recently launched the first of its operational *Meteor* weather satellites.

So far the emphasis has been on obtaining cloud pictures in both visible and infrared light, and temperatures of land, sea and cloud surfaces from rather simple radiometric measurements. The latest satellites, such as *Nimbus* 3 (Fig. 18), also carry sophisticated spectrometers to determine the vertical distributions of temperature and humidity throughout the whole depth of the atmosphere and so provide, for the first time, quantitative global data of the kind required for numerical weather prediction.

In this chapter I shall describe the requirements for global atmospheric measurements, how they are being obtained, and how they are likely to be obtained in the future by satellites carrying radiometric sensors and interrogation systems. It is likely that the global weather observing and communication systems will become largely based on satellite technology during the next 10–20 years.

Why global observations?

Each day of the year about 8000 land stations and 4000 merchant ships make regular observations of the weather as experienced at the Earth's surface. Conditions in the upper air are recorded at only a few hundred stations, which send up balloon-borne instruments

46

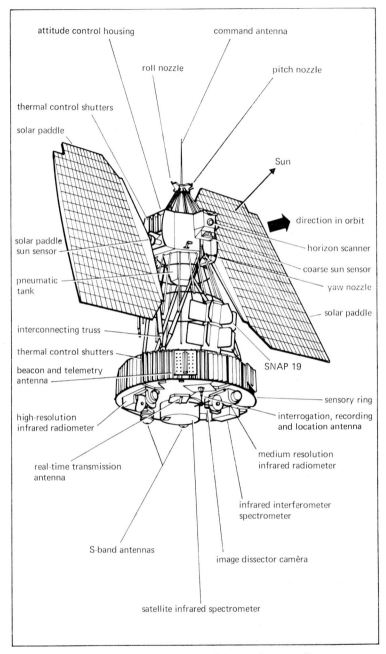

attitude control housing

command antenna

roll nozzle

pitch nozzle

thermal control shutters

solar paddle

Sun

direction in orbit

solar paddle
sun sensor

horizon scanner

coarse sun sensor

pneumatic
tank

yaw nozzle

solar paddle

interconnecting truss

thermal control shutters

SNAP 19

beacon and telemetry
antenna

sensory ring

high-resolution
infrared radiometer

interrogation, recording
and location antenna

medium resolution
infrared radiometer

real-time transmission
antenna

infrared interferometer
spectrometer

S-band antennas

image dissector camèra

satellite infrared spectrometer

FIG. 18   Basic configuration of the *Nimbus* 3 satellite.

called radiosondes. The instruments record the air temperature pressure and humidity as they ascend (and later descend on a parachute), and transmit their readings to the ground station or weather ship in the form of coded radio signals. The balloon also carries a reflector that permits its path to be tracked by radar, so that winds can be measured up to heights of 30 km (100 000 ft) or more. There are more than 600 radiosonde stations in the Northern Hemisphere, making soundings every 6 or 12 hours.

The density of the network is adequate over Europe and North America but too sparse elsewhere, especially over the oceans and the tropical regions. The Southern Hemisphere – largely ocean – has a total of less than a hundred radiosonde stations. In other words, the atmosphere over at least half of the globe is not observed adequately to describe and forecast even the broad features of the global circulation.

Nevertheless, recent research encourages the belief that, given adequate observations and computing facilities, it should be possible within the next few years to produce reliable forecasts of the main weather patterns for 5–7 days ahead, and a useful indication of general trends over periods of perhaps 2–3 weeks. Forecasts on these timescales will require very complex physico-mathematical models of the atmosphere, computers many times more powerful than existing machines, and adequate observations covering at least a hemisphere and perhaps the whole globe.

Additional observations will therefore be needed, especially from the oceanic and tropical areas and the Southern Hemisphere; also faster and more reliable methods of communication for the transmission and exchange of these data, together with much more powerful computers than exist at present. The main objective of the World Weather Watch (WWW) is to obtain just these facilities. Plans for the first four-year period (1969–72) call for the establishment of a minimal global network of observations by creating about 40 new fixed radiosonde stations and providing similar facilities on 100 moving ships, and for installing a high-speed, global-communication circuit linking the major meteorological centres, which will be equipped with powerful computers for processing the observational data and producing numerical forecasts. However, it is clear that an adequate global system cannot be produced at reasonable cost simply by expanding the current operational facilities. The World Weather Watch must ultimately depend largely on satellites for surveillance and remote sensing of the atmosphere, for the location and interrogation of instrumented balloons, ocean buoys and unmanned land stations, and for world-wide communication.

At this stage we cannot specify precisely the minimum and optimum standards for the coverage – the spatial resolution, frequency and accuracy of the observations, for example. This will be one of the main tasks of the Global Atmospheric Research Programme (GARP), the research component of the World Weather Watch. GARP proposes to undertake intensive observations of the global atmosphere over a period of one year in the late 1970s, with a view to determining the stability and limits of predictability of atmospheric behaviour, and the minimum observational and other facilities that will be necessary for the fully operational phases of the World Weather Watch. GARP also plans to investigate the evolution of the mesoscale tropical weather systems and their role in transporting heat and water vapour into the global circulation, and the exchange of heat, momentum and water vapour between the oceans and the overlying atmosphere.

So far as we can judge at present, it seems likely that, in order to forecast the large-scale dynamical features of the atmosphere that will determine the essential character of the weather everywhere for a week ahead, vertical soundings of temperature, pressure and wind up to heights of 30 km and of humidity up to 10 km will be required once, perhaps twice, daily at an average horizontal spacing of 500 km over the entire globe. A similar density of measurements of the basic surface paarameters, including sea-surface temperature, would also be required.

In terms of present facilities, this would imply an impossible requirement – about 1500 additional upper-air stations, mainly over the oceans. Only by using satellite techniques would adequate global coverage become possible.

Meanwhile, 5–7 day forecasts for at least the middle latitudes of the Northern Hemisphere might be possible if the present network of upper-air stations could be reinforced over the oceans, particularly at low latitudes, to give an average spacing in these regions of 1000 km. Northern Hemisphere forecasts for 10–20 days will probably require, in addition, a similar density of observations from the Southern Hemisphere. As for accuracy, air temperatures to within 1 °C, atmospheric pressures to within 0·3 per cent (3 mb at the surface), winds to within 1–2 m/s, vapour pressures to within 10 per cent and sea-surface temperatures to within 1 °C may prove adequate. But all these assumptions will need to be tested, first in long-term numerical simulation experiments on the computer, and ultimately in predictions using real initial data.

Satellites can provide efficient and economic global observations of meteorological factors by measuring the intensity, polarization and angular and spectral variation of radiation emitted and reflected by the Earth and the atmosphere. In polar orbit they have the great advantage of being able to survey a large area of the atmosphere at any one instant, and to cover the whole globe every 24 hours. Despite

49

their distance from the Earth, the difficulties arise not so much in finding instruments that will make radiation measurements with the desired accuracy, resolution and coverage, but rather in discovering and exploiting physical relationships that uniquely relate the measured quantities to the desired meteorological factors. As we shall see later, it is one thing to measure the spectral distribution of radiation received by the satellite from the underlying atmosphere, and quite another to deduce from these measurements the vertical distribution of atmospheric temperature or humidity with the accuracy required for use in numerical forecasting.

Cloud photography and mapping

Ever since the first remarkable pictures from *Tiros* I in 1960, the meteorologist has received an almost continuous flow of excellent photographs from a succession of US satellites. Since early 1966, the introduction of the automatic picture transmission facility has made these pictures available in real time to any country that can make, or buy for a few thousand pounds, the necessary receiving and reproduction equipment. About sixty countries are now so equipped, and the photographs showing the cloud formations, and the distribution of sea ice and snow on the ground, are proving a most valuable aid in weather forecasting.

Present techniques allow the television picture to be resolved into 1000 by 1000 individual points, which, on the current *Nimbus* and *Essa* satellites, corresponds to a resolution of about 1 km on the surface of the Earth. For continuous coverage of the Earth, this amounts to 10^9 data points per day for storage and transmission to ground stations.

Excellent pictures covering the eastern half of the North Atlantic, Western Europe and the Mediterranean are received daily at Bracknell (near London). They show very clearly the organization and structure of cloud systems ranging from large cyclonic vortexes a thousand miles across to individual shower clouds only one mile in diameter (see Fig. 19). Often, by comparing these photographs with the weather charts, a better analysis and forecast has been produced. Their detailed analysis and interpretation has helped to locate mesoscale features such as hurricanes, thunderstorms, squall lines, jet streams, mountain waves and regions of strong wind shear, often from oceanic and other remote areas where the observing networks are too sparse to make their detection likely by conventional methods. Clearly revealed, too, is the distribution of sea ice, and of land areas covered by ice and snow.

The intensity of the radiation emitted by the cloud, land and sea surfaces has been measured with scanning radiometers (photometers) on satellites orbiting at distances of the order of 1000 km. In order to ensure that the radiation sensors point continuously at the Earth the satellite has to be stabilized with an accuracy of about ±1 degree in three axes with respect to the centre of the Earth (see Chapter 1). The spatial resolution depends largely on the spectral band

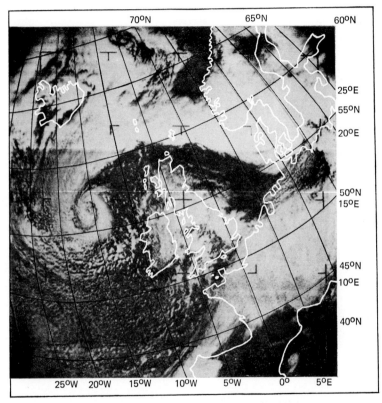

Fig. 19 Photograph from a US *Tiros* satellite showing the British Isles partially covered by cloud in the centre of the picture and a spiral cloud pattern associated with a small depression centred west of Ireland.

used, and is limited by the minimum acceptance angle of the detector for a given radiometric accuracy, and by the permissible size and scanning speed of the instrument.

Working in the infrared window near 11 μm, with angular resolutions of about 0·5 degree, *Nimbus* and the latest *Essa* satellites have provided continuous global mapping of cloud formations and the Earth's surface, by day and night, with a spatial resolution of 5–10 km, and have determined the equivalent black-body temperatures of the surfaces to within an accuracy of perhaps 2°K. In the next few years we may anticipate the use of more sensitive detectors able to achieve a resolution comparable to that achieved with television cameras in the visible spectrum.

A geosynchronous satellite – that is, one in equatorial orbit synchronized in period with the rotation of the Earth – remains stationary over a particular point on the equator, and therefore allows one

Fig. 20   Picture transmitted (in colour) from the US ATS–3 geosynchronous satellite over the Amazon, showing the whole of S. America, N. America and Western N. Africa. Prominent among the many cloud systems are the spiral pattern associated with a small depression off Morocco, a long line of convective clouds stretching east–west over the southern part of the North Atlantic, and a cold frontal system over Uruguay and North Argentina. (*Courtesy of NASA.*)

particular area to be observed continuously. The US ATS-1 satellite has in effect a 5-inch telescope which, by scanning laterally while the satellite spins about its axis, views a wide strip of the Earth every 20 min. In this way it provides a time-lapse film showing the development and decay of cloud systems over nearly one-quarter of the globe. Fig. 20 shows a picture taken from a second satellite poised over the Amazon, which views the whole of North and South America, Western Africa and most of the Atlantic Ocean. Four of these satellites would suffice to cover the whole globe (except for relatively small areas around the Poles). But the pictures from high latitudes are inevitably distorted because the viewing angle is very oblique. At present the remoteness of the geosynchronous orbit (36 000 km) restricts observations to the visible spectrum and therefore to daylight hours, but improvements in the design and operation of infrared detectors eventually should remove this limitation.

A satisfactory global coverage of cloud pictures might be achieved by using a combination of four geosynchronous and one or two polar-orbiting satellites. The geostationary satellites could relay pictures received from the polar-orbiting satellites to those points of the Earth not in direct view of the polar satellites.

Measurements of radiated energy

First, let us consider solar radiation, the primary source of energy for driving the atmospheric circulation, which can be measured accurately outside the atmosphere only by rocket or satellite. Measurements of the solar flux can be made either with high-resolution spectrometers on Sun-stabilized satellites, or with wider-band radiometers on Earth-oriented satellites by scanning the Sun for, say, 20 minutes during each orbit. At present, measurements of the solar constant have probable errors of at least ± 2 per cent, while the intensity of the ultraviolet portion is known only to within a factor of two. The outgoing solar radiation reflected back to space by the atmosphere and the Earth has been measured on American and Russian satellites, but there are difficulties in estimating the total hemispheric flux from narrow-beam measurements and in allowing for angular and spectral variations.

The outward flux of long-wave terrestrial radiation in the 4–30 μm band was one of the first measurements to be made from a satellite. Broadband radiometers can measure the intensity of radiation received from a given direction with an accuracy of 1 per cent, and with angular resolution of better than 1 degree. Since the filters and detectors used in radiometers do not permit uniform response over the entire spectral range, and because each area on Earth is viewed through a narrow aperture, it is again necessary to use analytical and empirical models to relate the measurements to the total emitted flux over the hemisphere.

Nevertheless, sufficiently accurate measurements of both incoming and outgoing radiation have been made to derive useful global maps of the net amount of energy available to the Earth and atmosphere, and to give some indication of the day-to-day changes in the heat balance of the system.

If a homogeneous surface radiates uniformly and isotropically as a

53

black body, its temperature can be deduced from a measurement of the radiation received in a given spectral range, provided that the intervening medium is transparent to these particular wavelengths. Thus satellite-borne radiometers, working in the 3·2–4·2 μm and 10·5–12 μm atmospheric windows, have determined the surface temperatures of cloud layers and of the oceans (both of which act as good black bodies) to within about 2°K. Measurements from polar-orbiting satellites have been compiled to produce maps of cloud-

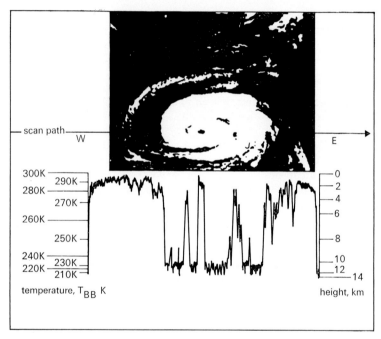

FIG. 21   Infrared photograph of a hurricane taken from a Nimbus satellite. The temperatures of the ocean or cloud surfaces as determined from measurements of emitted radiation are shown by the trace below the photograph.

summit temperatures and of temperature anomalies in the oceans (see Fig. 21), although continuous tracking of the oceans is often prevented by intervening cloud.

In the case of non-raining clouds at least, interference could largely be overcome by using microwaves, for example in the 1·6 cm window. But many natural surfaces are not black for microwaves, and their emissivities would have to be determined. This fact might be turned to advantage, however, in distinguishing sea water (emissivity $\varepsilon = 0·3$) from ice ($\varepsilon = 0·9$), even through haze and fog.

Cloud pictures and radiation data from satellites currently provide a great deal of valuable information. Unfortunately it is mainly qualitative in character and very difficult to incorporate directly into numerical weather-prediction models. These models require data on atmospheric temperature, pressure, composition and winds. What, then, are the possibilities of obtaining such information by remote sensing of the atmosphere from satellites?

Remote sensing of temperature

In contrast to the simple radiometers used to measure total fluxes and surface temperatures, some very sophisticated and ingenious spectrometers are being developed to obtain the vertical distribution of atmospheric temperature. The idea is to measure the spectral distribution of radiation received from constant constituents such as carbon dioxide or molecular oxygen.

The radiation emitted by a layer of uniformly mixed gas in local thermodynamic equilibrium is directly dependent on its temperature. But the intensity of the infrared radiation reaching a satellite from, say, the carbon dioxide distributed throughout the depth of the atmosphere, even through a window in which absorption by other constituents is negligible, will also be affected by absorption within the gas itself. Relatively little radiation would be received from the comparatively small mass of emitting gas in the uppermost layers; and relatively little from the lowest layers, whose emission is strongly attenuated in traversing the whole depth of the atmosphere.

Radiation received therefore comes mainly from the intermediate levels, and must be weighted according to height (or total pressure), as indicated by a function of the form shown in Fig. 22(a). The half-width (δ) of this curve is typically a few kilometres so that, in effect, the received radiation is a measure of the mean temperature of a layer of atmosphere of this thickness. Moreover, since the absorption is strongly dependent on wavelength, most of the radiation transmitted in the strongly absorbing wavelengths will reach the satellite from the upper regions of the atmosphere nearest the satellite. For less attenuating wavelengths, the received radiation will come from progressively lower levels. In other words, the transmittance τ between any level Z and the top of the atmosphere is a strongly weighted function of wavelength λ and Z.

The intensity of radiation received by the satellite within a very narrow band of central frequency may be written

$$I(v) = \int B\{v, T(Z)\}\frac{d\tau(v, Z)}{dZ}.dZ$$

where B is the Planck (black-body) radiance and T is the temperature. The weighting function $d\tau/dZ$ reaches maximum values at

55

different levels according to the frequency (or wavelength), as shown in Fig. 22(b).

Thus a set of measurements of the radiation received at different wavelengths contains information on the vertical distribution of temperature in the atmosphere. In effect, the received radiation gives information on the mean temperatures of a number of overlapping

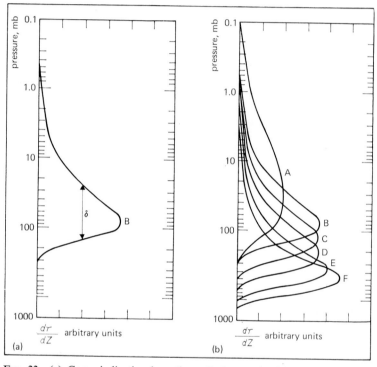

(a)

(b)

FIG. 22   (a) Curve indicating how the radiation received by a satellite from the carbon dioxide in the underlying atmosphere is weighted according to height because of absorption of the emitted radiation within the gas itself. (b) Weighting functions drawn for six different wavelengths in the 15 μm infrared band. Showing how the transmittance is weighted towards lower levels in the atmosphere for less strongly absorbing wavelengths.

atmospheric layers. If the weighting functions did not overlap, it would be a simple matter to derive the temperature profile from the radiation measurements. But they do overlap, so the individual measurements are highly interdependent and have to be made with great accuracy. The mathematical techniques for inversion of the set of radiative transfer equations have to be applied with great care and ingenuity to obtain unambiguous and stable solutions.

Various techniques are being developed for selecting the narrow frequency intervals and measuring the radiation received in them. The *Nimbus* 3 satellite, launched by the US in April 1969, carries an Ebert infrared spectrometer (SIRS). It uses a fixed grating to select seven narrow spectral intervals of width about 5 cm^{-1} in the 15 μm carbon dioxide band, and one interval in the 11 μm window. A 'chopper' switches the sensor between the radiation received from the underlying atmosphere and from a black body of liquid nitrogen temperature, to produce an alternating signal which is detected by a set of thermistor bolometers. The satellite also carries an infrared interferometer spectrometer (IRIS) – in essence, a Michelson interferometer – to accomplish the same task by scanning the entire spectrum from 6 to 20 μm with a resolution of about 5 cm^{-1}.

A simpler device, having several advantages, has been developed at Oxford and Reading universities and was launched on the *Nimbus* 4 satellite in April 1970. It employs narrow-band (3·5 cm^{-1}) interference filters to select six channels in the 15 μm band, together with absorption cells containing carbon dioxide to filter and select further the required radiation. The detectors are bolometers. The carbon dioxide filter removes radiation at wavelengths corresponding to the centres of the absorption lines (mainly from the upper layers), which sharpens the weighting function and biases it towards lower levels where improved resolution is now possible. Alternate switching between a carbon dioxide-filled cell and an empty cell gives two sets of readings, whose differences may be used to obtain improved information also at higher levels and extend the measurements of temperature up to 50 km. The equipment weighs only 10 kg and uses only 5 W of power.

The performance of SIRS, the spectrometer aboard *Nimbus* 3, has exceeded expectations in providing excellent vertical temperature profiles, especially over the Northern Hemisphere, in areas free of clouds.

At present, retrieval of a temperature sounding from the eight discrete radiance measurements is not accomplished by direct inversion of the radiative transfer equations, but by relating the measured radiances to atmospheric temperatures through statistical regression equations derived from a large sample of radiosonde soundings. Regression equations – up-dated every few days – are determined for five different latitude belts between the North and South Poles. Equations are derived for both the temperature and the geopotential height of 13 standard pressure levels between the surface and 10 mb (30 km). Intermediate levels are computed from the hydrostatic equation (given on page 61), so that each profile is determined by 25 levels. The radiances are measured over a field of view of approximately 225 km square, so the temperature profiles represent an average over this area. The eight channels are sampled simultaneously at 8-second intervals, during which the spacecraft travels about 50 km. Global coverage is achieved twice each day – at noon and midnight local time. Day-time and night-time observations are obtained with equal precision.

Presence of clouds in the field of view causes a serious problem in determining a proper temperature sounding below the tops of the clouds. With a heavy cloud cover, the sounding from the cloud top to the surface must be interpolated between the satellite-derived temperature and a known surface air temperature. With a partial cloud cover, the radiance measurements are affected by the cloud amount and the heights of cloud layers.

In this case, a statistical technique is used to derive a simulated clear-air sounding that SIRS would have obtained had the clouds not been present in the field of view. Data from the more opaque SIRS channels (which measure radiance mainly from higher levels and are unaffected by tropospheric clouds) are used with the surface air temperature to obtain a first approximation to the clear-air sounding below the clouds. Data from the remaining channels are compared with the radiance values implied by this first approximation. The departures are used, with several iterative steps, to improve the approximation to the clear-air sounding in the lower levels, and to produce an estimate of height and amount of one or two cloud layers which would account for the observed departures.

More than 90 per cent of the SIRS soundings now require some adjustment because of the presence of clouds. Future sounding instruments being developed for the operational satellite system will have a smaller field of view, of the order of 50 km square, and will scan a much broader swath perpendicular to the orbital track. These improvements will greatly increase the number of soundings obtained in clear-air regions between cloud masses.

Fig. 23 shows the historic first temperature sounding derived from satellite measurements and, for comparison, the radiosonde sounding taken from Kingston, Jamaica, about 400 km to the north-west and four hours earlier in time. Close agreement between the two is evident. The main difference lies in the middle troposphere, where the SIRS sounding is warmer by 2° or 3°C. Systematic errors of this type are rather typical and arise mainly from inadequate knowledge of the transmittance functions of carbon dioxide and water vapour. In practice, they may be reduced by using past SIRS observations coincident in space and time with radiosonde data as observed *radiances* in deriving the statistical regression equations, thus avoiding the need to specify the carbon dioxide transmittances explicitly.

When these rather empirical methods are used to deduce the temperature profiles, the discrepancies between the SIRS-derived and radiosonde observed temperatures are between 1·5° and 2·0°C for 70 per cent of soundings in the Northern Hemisphere. Less than 2 per cent of the discrepancies exceed 6°C, the largest errors occurring in the lower troposphere due to the influence of clouds. The data for the Southern Hemisphere are not so good because there are so few radiosonde soundings on which to base the regression equations;

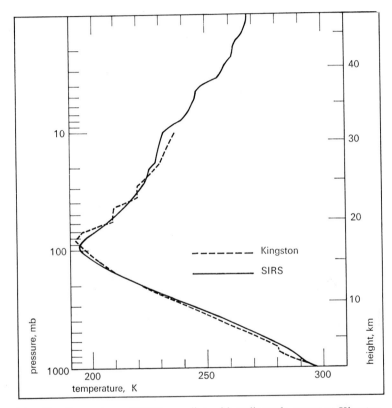

FIG. 23 Comparison of SIRS sounding with radiosonde ascent at Kingston, Jamaica, 14 April 1969, under clear-sky conditions.

here we may have to wait for satisfactory methods of inverting the radiative transfer equations.

Within a few weeks of obtaining the first temperature profiles, the SIRS data were being incorporated in real-time analysis of the temperature, pressure and geopotential fields by the United States Weather Bureau, and they have already proved of considerable value in filling major gaps in the conventional observations. As an example, Fig. 24 shows analyses of the geopotential heights of a 500-mb

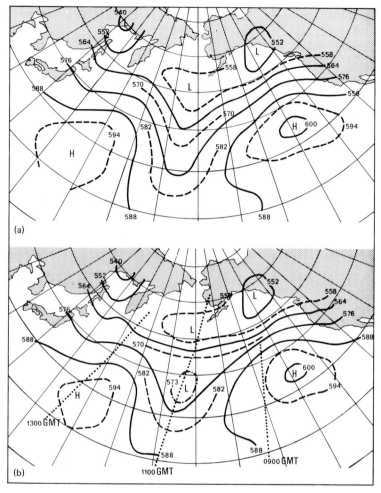

FIG. 24   Analyses of the 500-mb geopotential field (a) without and (b) with
SIRS data, 0000 GMT, 24 June 1969.

pressure surface, with and without the SIRS data. The satellite information at 1100 GMT indicated that the deep trough in the central
Pacific contained a cut-off, low-pressure centre which was not
detected by the conventional analysis because of the sparsity of
radiosondes in this area.

The experiments on *Nimbus* 3 and *Nimbus* 4 have demonstrated
convincingly that atmospheric soundings from satellites are feasible.
With further improvements in instrumentation and analysis – and

several are already under development – it should be possible to produce a global coverage of observational data at least as accurate as that now being provided for limited areas by conventional methods. In the stratosphere the SIRS data are probably already more reliable than radiosonde data, which are subject to large instrumental errors at these levels.

Measuring atmospheric pressure or density

Accurate determination of the vertical temperature distribution would allow the pressure (or density) profile to be calculated from the hydrostatic equation:

$$\frac{1}{p}\frac{\delta p}{\delta Z} = \frac{-g}{RT}$$

provided that the surface pressure p_0 were known. Accurate measurement of the surface pressure on land is readily achieved by modern automatically-recording aneroid barometers. The problem is much more difficult over the oceans where it would be possible – if very expensive – to mount aneroid barometers on floating buoys that could be located and interrogated by satellites. However, this may be rendered unnecessary by the recent development, in the USA, of a simple, lightweight radar altimeter which, when mounted on a constant-level balloon at, say, 100 mb ($\sim$16 km), can measure the height Z of the balloon above the sea-surface to within an accuracy of ± 10 m. Hence if the balloon carries instruments to measure the ambient pressure and temperature, the parameters p, T and Z in the hydrostatic equation are known, and p_0 may be calculated. Given now the surface pressure, and T as a function of Z from the infrared spectrometer data, the pressure may be determined as a function of height Z from the hydrostatic equation.

Alternatively the profile of atmospheric density could, in principle, be obtained directly from satellite measurements of the refraction of starlight, the refraction of microwaves or the absorption of light by the molecular constituents of the atmosphere. For example, the satellite could make continuous measurements of the angular position of a given star, which would depend on the degree of refraction the light rays suffer in passing through the Earth's atmosphere.

A series of such measurements, made between the time when the star is acquired slightly above the horizon until occultation occurs, that is over a period of about 30 seconds, could be used to derive a profile of refractive index (and hence air density) at a point where the rays are tangential to the Earth – on the assumption that the atmosphere is spherically stratified. The method has not yet been tested, but calculations suggest that it should be possible to map the profiles in the lower stratosphere and upper troposphere with an accuracy of a few per cent. It will probably not be possible to make measurements at lower levels – say, below 5 km – because of atmospheric extinction and occultation by cloud.

It has also been suggested that, since the velocity and direction of propagation of microwaves are functions of the refractive index, and hence of the density, of the atmosphere, their precise measurement could be used to determine atmospheric density and its variations. The technique envisages a master satellite followed by five or six 'slaves' in low orbit around the Earth. Microwaves transmitted between the master and the slaves would traverse the atmosphere at five or six different levels, and accurate measurement of their velocity and phase would continuously measure the bending and retardation of the waves averaged over the horizontal ray paths at these levels. Measurement of a 1-mb pressure change would require the system to detect a change of 1·4 m in the apparent position of two satellites. This should be possible by Doppler measurements provided the physical separation of the satellites were known with great precision. In a dry atmosphere it might be possible to derive the pressure field to within ± 1 mb in the horizontal and within ± 5 mb in the vertical. The presence of water vapour, however, may make the results less accurate, particularly at low levels.

The intensity of the radiation reflected from the Earth's surface on arriving at a satellite would depend on the total quantity of absorbing gas in its path; that is, on the total pressure. This suggests that the total (surface) atmospheric pressure could be obtained by observing the intensity of sunlight or a strong laser beam reflected from the Earth's surface in the region of the visible absorption bands of oxygen near 7600 Å. Again, however, the method would be adversely affected by cloud.

Humidity measurements from satellite

In principle, the inversion methods developed for obtaining the temperature profile from measurements of radiation emitted by carbon dioxide (whose concentration is known) could also be used to obtain the (unknown) vertical distribution of water vapour, provided the temperature profile were known and corrections were made for the radiation from underlying cloud or land surfaces. With these provisos, it should be possible to obtain the weighted average of the relative humidity of the upper troposphere from measurements of emitted radiation in the 6·3 μm band and in a window between 8 and 12 μm; and by making additional measurements in the broad band between 19·5 and 24 μm, it should be possible to deduce the mean humidity in the lower and middle troposphere as well. The SIRS instrument carried on the *Nimbus* 4 satellite includes six additional channels for this purpose.

The accuracy of all such methods is limited by the fact that all particles in the atmosphere absorb and emit in the infrared. This suggests the use of the microwave emission at 1·35 cm. But at these wavelengths the emissivity of the underlying land surface is so great that the method seems feasible only above the oceans.

Measurement of global winds

No really feasible method has been proposed for sensing winds simultaneously at a number of levels in the atmosphere from a satellite. Determination of the pressure and temperature fields, as described above, would allow the winds to be calculated in middle and high latitudes where the winds are closely related to the pressure field through the controlling action of the Coriolis force due to the Earth's rotation. This would not be possible in the tropical regions, where the horizontal gradients of temperature and pressure and the Coriolis forces are all weak. Elsewhere, however, it may well prove possible to derive the large-scale wind field from measurements of surface pressure and vertical temperature distribution only.

Meanwhile, US meteorologists are devoting considerable effort to the extraction of winds from the movement of clouds on the pictures transmitted on successive frames from the spin-scan cameras carried on the geostationary ATS satellites. Although this technique does not allow us to obtain a complete horizontal coverage of winds, or information at more than about two levels in the vertical, it is producing much useful information in data-sparse areas such as the tropics where there are very few other sources of wind data.

The greatest attention, however, has been paid to the measurement of winds by tracking inextensible, constant-volume, super-pressurized balloons that drift with the winds along surfaces of constant atmospheric density. In the US GHOST (Global Horizontal Sounding Technique) Project, more than 200 such balloons have been released from Christchurch, New Zealand, and tracked by the signals from a simple telemetering photoresistor device that measures the elevation of the sun. Balloons flying at 100 mb and higher are able to remain aloft for many months and make many circuits of the globe. One has flown for more than 420 days, circling the Southern Hemisphere more than 25 times at the 100-mb level. Balloons flying at the 200-mb level (40 000 ft) have an average life of about 100 days.

Fig. 25 shows the trajectories followed by one such balloon in making eight circuits of the Southern Hemisphere over 102 days. Unfortunately balloons drifting at lower levels suffer with varying severity from the accumulation of frost, ice and snow, which limits their duration to weeks, or even a few days. If the balloon life-time can be increased, the icing problem overcome, and an electronics package designed that will present no hazard to aircraft, this technique is likely to become an important component of the World Weather Watch. Adequate global data on winds and much valuable temperature and pressure data could be supplied by keeping perhaps

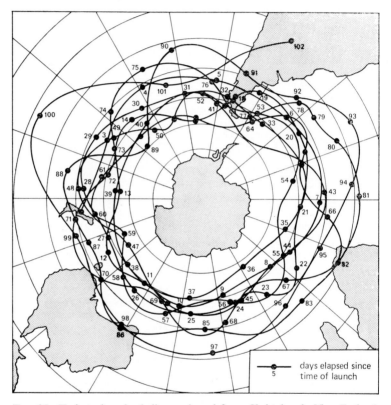

FIG. 25   Trajectories of a balloon released from Christchurch, New Zealand, making eight circuits at 200 mb (40000 ft) during 102 days.

6000 balloons circling the earth at altitudes between 3 and 25 km. They would be located and interrogated by a system of satellites and their data relayed to a ground-based computer for analysis.

Three alternative systems are being investigated (see Fig. 26). The IRLS (Interrogation, Recording and Location) System, under development by NASA, will determine the location of a balloon by measuring the propagation time of a two-way radio signal between the satellite and the platform for two positions of the satellite in the

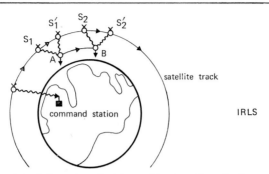

A is the position of the balloon determined by satellite at positions S_1, S_1'.
B is its position determined by satellite at positions S_2, S_2' in next orbit

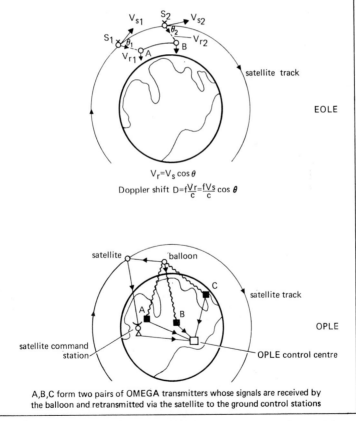

$$V_r = V_s \cos\theta$$

Doppler shift $D = f\dfrac{V_r}{c} = \dfrac{fV_s}{c} \cos\theta$

A,B,C form two pairs of OMEGA transmitters whose signals are received by
the balloon and retransmitted via the satellite to the ground control stations

FIG. 26 Diagrams illustrating the principles of the three location and interrogation systems – IRLS, OPLE and EOLE – of tracking balloons by
a satellite.

same orbit. These two range measurements and the exact time of the measurement in relation to the known geometry of the orbit are all that is required to establish the position of the balloon. A similar determination made during the next orbit, about 100 minutes later, will give the balloon's displacement and hence the average wind over this time interval. Design studies suggest that the total error in range measurement may not exceed 0·5 km, in which case it should be possible to locate a balloon to within ± 3 km and give average winds within ± 1 m/sec.

The EOLE system, designed by the Centre National d'Études Spatiales of France, will measure the Doppler shift of a signal transmitted from a satellite to the balloon and back again, and hence the angle between the propagation path and the tangent to the satellite orbit. Again measurements at two positions of the satellite fix the position of the balloon, and similar measurements during the next orbit yield the wind averaged over the intervening period. The first major trial, involving 500 constant-level balloons which will carry small thermistor thermometers and lightweight aneroid capsules, is scheduled for 1971. A target of ± 4 km has been set for the accuracy of location of a balloon.

The OPLE (Omega Position Locating Experiment) System, designed by NASA, makes use of the VLF (10 kHz) Omega navigational system which, because of its low attenuation and high phase stability, affords world-wide coverage and very accurate phase measurements. The balloon would receive signals from two pairs – that is, three – Omega ground stations and re-transmit them to a satellite which would then relay them to a computer in the ground for calculation of the balloon's position. Two lines of position (hyperbolic isophase contours) are established by the phase differences between the signals from each of the two transmitter pairs and the position of the receiver (balloon) is given by the intersection of these two contours. The very long base lines between the ground stations result in the position contours cutting nearly at right-angles, to give a fix that is expected to be accurate to within 1 mile by day and within 2 miles at night for almost all geographical locations.

Balloon and satellite need carry only a simple transponder. As the motion of the satellite – which acts only as a relay station – plays no part in determining the fix, it can be geosynchronous. By interrogating on 50 different frequencies simultaneously it might be possible for one geostationary satellite to interrogate a total of 2000 balloons for 3 minutes apiece every two hours. Three such satellites, one with its orbit inclined at 30 degrees to the equator, could achieve global coverage over each 24-hour period and interrogate a total of perhaps 6000 balloons.

However, because of the severity of the icing problem, to which no solution has yet been found, balloons can be flown economically only above 200 mb, and below 850 in the tropics, the most vital part of the troposphere thus being excluded. Accordingly, it seems that the constant-level balloon will play a much less prominent role in the World Weather Watch than once seemed likely. This leaves unsolved the difficult problem of obtaining wind data in the tropics. But it has been suggested that a large, constant-level balloon flying at, say, 100 mb, might release very lightweight drop-sondes carrying transponders and fitted with parachutes that would drift with the wind. These transponders could be located and tracked by a navigational system such as OMEGA. Such a system may be tested within the next year or so.

Automatic weather stations

All three of the satellite systems I have just described are capable in principle of communicating with automatic weather stations on land, on anchored or floating ocean buoys, and on moving merchant ships. In all cases, identification and precise location will be easier than with balloons.

Several countries are developing automatic stations to report conventional surface weather observations. In Britain, the Meteorological Office is about to put into operation stations that measure pressure, temperature, humidity, windspeed and direction, rainfall, visibility and sunshine. They can be interrogated either by radio or a telephone call. Similar installations will also be tried out on merchant ships, where the tendency towards greater automation and smaller crews may make it increasingly difficult to obtain all the required information from voluntary observers. The satellite interrogation of automatic ship-borne stations may therefore become an important feature of the World Weather Watch.

Even so, there will be vast areas of the oceans almost devoid of ships, and here automatic stations on buoys are a possible solution. However, the technical problems of making sensing and telemetering systems that will continue to work unattended for several months on the high seas, and to do this at a cost which will allow them to be deployed in large numbers, are formidable. So major expenditure on buoy development is unlikely to materialize until other and cheaper solutions have been thoroughly investigated.

The future

There is little doubt that the developments I have described in this Chapter will, over the next decade or two, lead to major changes in the methods and techniques of meteorological observation and measurement. Direct measurement of some of the basic atmospheric parameters will be gradually supplemented, and perhaps ultimately replaced, by indirect methods involving the remote sensing of electromagnetic radiations from atmospheric constituents by satellites. For the next ten years or so, the global observing system will probably consist of a mixture of satellites, conventional radio-sondes, dropsondes, balloons released from ground stations and ships and tracked by navigational aids such as LORAN and OMEGA, constant-level balloons carrying accurate radar altimeters, and so on.

Satellite methods, however, are likely to become increasingly dominant. Although it may be necessary to retain some ground-based sounding stations for the calibration and checking of the satellite systems, it will probably be possible to dispense with many of the radio-sonde stations in the Northern Hemisphere and the weather ships well before the end of the century.

Satellites will also play an important role in locating and interrogating automatic instruments on remote land stations, ocean buoys and drifting balloons, and in rapidly transmitting the great quantities of data to a few major centres for processing and analysis by giant computers. An adequate global system would consist of four geostationary satellites, equally-spaced around the equator, and two polar-orbiting vehicles. Such a system is likely to be in operation before the end of the present decade.

CHAPTER **4**

Earth Resources Satellites

Watson Laing
Hawker Siddeley Dynamics

Remote sensing has been associated with aerospace activities since
the first crude aerial photography from balloons. Reconnaissance
devices have evolved from the simple cameras carried in First World
War biplanes to the sophisticated scanners, detecting and recording
radiation beyond the visible spectrum, carried by present-day
reconnaissance aircraft. In the US, for instance, very advanced types
of camera were developed for high-flying aircraft of the U2 type and
later for 'spy-sats' in the *Samos* and *Discoverer* series.

The phrase 'Earth resources' has now come to mean not only
minerals and oil, but also water supplies, crops, forests, fisheries and
pollution: every aspect, in fact, of Man's environment which can be
sensed or monitored. Present proposals for Earth resources satellites
describe stable, Earth-pointing platforms in circular orbits of a few
hundred miles altitude. These platforms would carry sensors to detect
electromagnetic radiation from one small area of the Earth's surface
at a time. The radiation may be reflected sunlight or energy radiated
by the Earth itself. It need not necessarily be visible; nor need the
information be composed in the form of images. The data from these
sensors would be returned by telemetry in digital form and processed
on the ground either into high-quality photographic images or des-
criptive numbers.

Many terrestrial features can be described by the way they selec-
tively emit or reflect radiation of given wavelengths, so providing
each feature with a spectral 'signature'. In size, and in structural and
thermal design, the Earth resources satellites will be very similar to
present low-altitude meteorological satellites, described in the pre-
ceeding Chapter. In general however, they must maintain more stable
attitudes so that blurring does not take place, and be more accurate in
reporting these attitudes to make certain that the area under examination is precisely defined. The Earth resources satellite also requires a

data handling sub-system of much greater capacity and speed than a meteorological satellite, since much smaller areas of surface are of individual interest; consequently, for the same overall coverage, their number is increased by the second power. Moreover, much more information than, say, cloud photography requires must be culled from each of these patches.

TABLE 4. Theoretical resolving powers at 142 miles altitude.

Focal length (inch)	Scale factor	Negative resolution (lines/mm)	
		40	100
		Ground resolution	
12	750 000	60 feet	24 feet
36	250 000	20 feet	8 feet
120	75 000	6 feet	2·4 feet

Information on spy satellites so far released shows they have carried very high resolution film cameras (see Table 4). After securing the required exposures the camera package is ejected, re-enters the atmosphere and descends on parachutes. The use of spy satellites is rarely admitted, and resulting photographs have – understandably enough – never been exhibited. Unquestionably they would be most valuable for Earth-resource analysis and could be used at once if released. The possibility of this happening in the short term is remote, although they will provide future historians with a high-resolution synoptic record of the Earth and its changes. More recent surveillance satellites have been of a more permanent nature, and are for warning of nuclear explosions and detecting the launching of missiles by means of their characteristic exhaust signatures (see Chapter 7).

By contrast, most Earth resources satellites will employ television cameras, will have lifetimes measured in years and will be non-recoverable (although proposals for some satellites include recoverable camera flights). The resolution of the television pictures for normal scene contrast will be greater than 100 feet – very poor by 'spy-sat' standards. The greatest difference, however, will be in the open dissemination of information. The US Government and its National Aeronautics and Space Administration (NASA) have adopted an extremely open and generous position over their Earth resources activities. In 1969 President Nixon issued an invitation to the nations of the world to participate in the US programme.

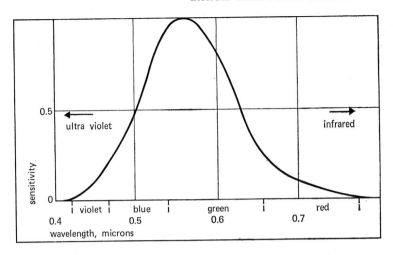

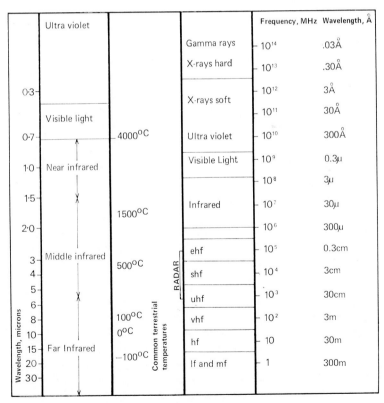

FIG. 27 Electromagnetic spectrum.

Remote sensing

The most familiar remote sensing instrument is the combination of human eye and brain. The eye operates as a frequency- and intensity-sensitive electromagnetic wave detector to radiation of a wavelength between 4500 and 7200 angstroms. The eye is linked to a high-speed, real-time data processor, and the combination can discern a very large number of intensity levels and perform rapid data compression and presentation. But its limitations include the eye's inability to detect polarization of incoming radiation, and the narrow bandwidth it senses. Energy absorbed beyond 7200 angstroms is detected only crudely, as heat, and only if sufficiently intense. Radiation entering the eye from 4500 down to about 3000 angstroms can be 'seen' as a grey colour but is physiologically harmful. The extent of this restriction is shown in Fig. 27, where the 'visible' band is shown in relation to the rest of the electromagnetic spectrum.

All hot bodies emit electromagnetic radiation in accordance with Planck's Law. The Wien displacement law is

$$\lambda_{max} \, T = \text{constant (2897 micrometre K)} \qquad (4.1)$$

Where λ_{max} is the peak wavelength of the Planck distribution curve, and T the absolute temperature. These show that the hotter the body the shorter the wavelength of its peak emitted radiation. The Earth is a hot body at a temperature of approximately 300 K as measured at the surface, or 270 K as observed from space (due to the generally lower temperatures of clouds). Space itself presents a background (stellar) temperature of about 4 K. The Earth acts as any other hot body and gives off total radiation in accordance with the Stefan–Boltzmann law:

$$W = \varepsilon\sigma T^4 \qquad (4.2)$$

Equation (4.1) indicates that the greatest intensity of radiation from the Earth must be emitted at about 10 micrometres – in the thermal infrared. This is our common experience when, for example, we pass close to sunlit brickwork. In reality the Plank curve is severely modified, especially over small Earth surface areas, since few real bodies have emissivities approaching one, and therefore are more grey bodies than black bodies. In particular, many surfaces and substances have emissivities which vary with both temperature and frequency. Fig. 28 gives a schematic impression of how the rate of energy given out by quartz at one temperature is distributed by wavelength. This phenomenon is known as the *reststrahlen* or residual ray effect and is a property of silicates. Fig. 29 shows how in general basic rocks can be distinguished from acidic rocks by signature since the exact wavelength is modified by the metallic ion content of minerals.

Another emissive peculiarity, exploited in camouflage detection, for example, is the chlorophyll absorption effect of healthy foliage. It is invisible, for it occurs in the near-infrared, but can be made visible by photographing with a film whose sensitivity extends beyond that of normal camera film, into the near-infrared. Commonly the photographs of this type may be printed in false colours to enhance the effect. We 'see' in the 'visible' at 4500 to 7200 angstroms because our eyes have evolved for daylight existence. The

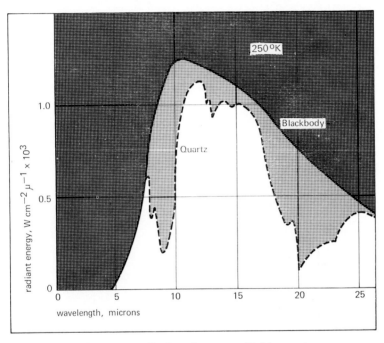

FIG. 28   Distribution of energy emitted by quartz.

characteristics of the Sun give our surroundings maximum reflectances in the visible band. There is no fundamental significance in these wavelengths; for example, many moths have developed dielectric microwave 'aerials' for night vision by emitted radiation.

Obvious questions arise as to whether there is any advantage or value in sensing or imaging at wavelengths other than the visible. Before answering these questions we must consider the effect of the atmosphere on the propagation of various radiations. The atmosphere in typical clear weather conditions has specific spectral 'windows' in which radiation is transmitted. Other frequencies are absorbed or scattered by atmospheric constituents, and to these frequencies the

73

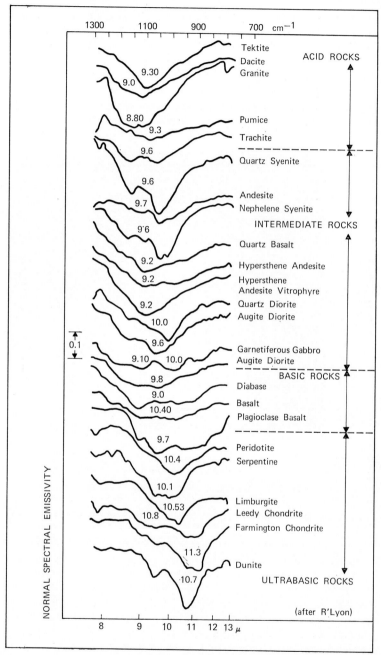

FIG. 29 Spectral signature of rocks.

atmosphere is opaque. The selective absorption varies with path
length, altitude and atmospheric conditions. In clear weather, we
can find some important 'windows', such as the 3·4 to 4·2 micrometre

TABLE 5.   Common microwave regions.

Designation (band)	Nominal wavelength (cm)	Nominal frequency (GHz)
L	25	1
S	10	3
C	6	5
J	4·5	6
X	3	10
K	1·2	25
Q	0·8	38

and 8 to 14 micrometre infrared windows, microwave and radio
windows and, of course, the visible band. Microwave (Table 5) and
radiowave frequencies have a further advantage in that they will
penetrate cloud.

TABLE 6.   Examples of passive and active instruments in the same spectral
regions.

Waveband (cm)	(working units)	Passive	Active
5×10^{-5}	5000 Å	Film camera with filter	Optical laser radar (Lidar)
1×10^{-3}	10 μ	Infrared line-scan Infrared radiometer	Thermal (CO_2) laser radar Laser altimeter
1	1 cm	Microwave imager Microwave radio-meter	SLAR Radar scatterometer

Remote sensing devices may be passive or active (see Table 6).
An active sensor contains the source of the energy to which it is
sensitive. An ordinary camera is a passive device, since it images and

75

records solar energy reflected from the scene photographed. Radar, on the other hand, is active since its basis is the transmission of electromagnetic pulses and the measurement of the signals reflected from the target. Sensors which can detect emitted radiation such as infrared or microwave scanners and those which are active are clearly independent of daylight and operate equally well in complete darkness.

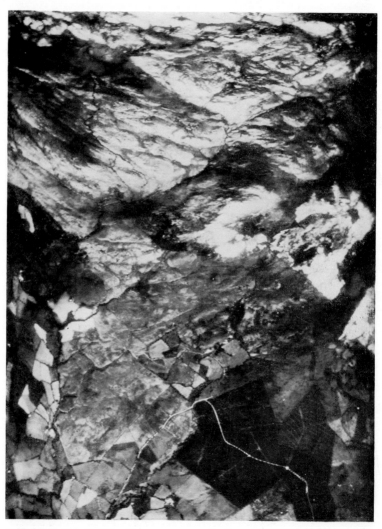

Fig. 30   Infrared linescan imagery. (*By permission, The Director, Royal Radar Establishment: Crown Copyright.*)

Again, information from remote sensors can be displayed in image or numerical form. I have already described the eye-brain combination as a narrow-band instrument for imaging. The eye effectively acts as a filter to the signals coming in. If we wish to visualize other wavelengths, we must convert these to visible images. For example,

FIG. 31   The Malvern Hills photographed by SLAR imagery. (*By permission, the Director, Royal Radar Establishment: Crown Copyright.*)

thermal infrared linescanners, which can provide pictures of effective ground temperatures, have a crystal sensitive to infrared radiation. Radiation from the ground is focused on to the crystal by a mechanical scanning system, and thereby converted to current or voltage changes, which when amplified can be displayed on an oscilloscope or used to expose film. When the scanner is aboard a moving platform, such as an aircraft or satellite, a linear image is produced on the film, representing a band on the ground across the line of flight. Film speed is synchronized with the speed of the platform so that these bands abut and a complete image of the overflown terrain is built up.

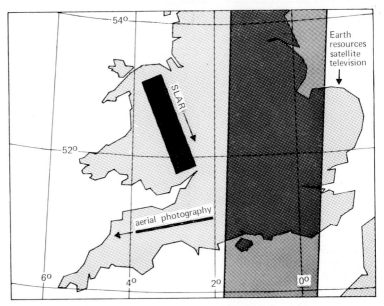

FIG. 32   Widths of swaths imaged by aerial photography, SLAR and satellite TV.

Sideways looking radar (in airborne form referred to as SLAR) also has an electronic signal as the primary output, which is usually used to produce photographic hard copy. Fig. 30 and 31 are examples of infrared linescan and SLAR imagery from low altitude. The large coverage and apparently high altitude of the SLAR should be noted. It has also been found useful to compare photographs taken in different parts of the visible spectrum using cameras and filters. This technique is termed multi-spectral photography.

It is also useful to compare photographs taken at different times to monitor the movement of fish shoals, icebergs or the progress of forestry operations. Fig. 32 gives an impression of the width of swaths imaged by aerial photography, SLAR, and proposed Earth resource satellite television cameras. Airborne infrared linescan is normally used at altitudes and with fields of view that give an even narrower coverage than aerial photography.

History

In April 1960, US scientists working on black-and-white visible-band TV cloud picture data from the newly launched *Tiros* satellite (see Chapter 3) observed ice in pictures of the Gulf of St Lawrence. When, in 1962, *Tiros* 4 was flown, the observation was exploited in the *Tirec* (Tiros Ice Reconnaissance) programme. Narrow-angle pictures were used to monitor the type, extent and concentration of ice in the Gulf, and to locate leads in the ice where ice-breakers might most easily force a passage. Ice surveillance has now evolved into a routine operational method.

US meteorological satellites in the *Tiros*, *Nimbus* and *Essa* series gradually established the credibility of Earth resource observations from orbital altitudes, despite the comparatively low resolution of the cloud sensors used. The most commonly reported types of observation included snow cover, broad geologic features, geothermal activity, ocean currents and inland hydrological systems.

High-altitude photographs of the Earth had been obtained from experimental aircraft and sounding rockets, but the interest of geo-scientists largely dates from pictures taken by an automatic camera aboard the first orbital Mercury spacecraft during the unmanned flight of September 1961. Successive manned spaceflights have returned increasingly valuable Earth photographs, culminating in the Apollo 9 multispectral terrain photography experiment. By 1965 NASA had entered into significant agreements with several other US Government and other agencies, including the National Academy of Sciences and the US Geological Survey, to develop, investigate and publicize remote sensing for both aerial and orbital investigations. NASA had also begun an aircraft experimental programme for instruments at its Manned Spacecraft Center.

In 1966 NASA launched its Earth Resources Technology Satellite (ERTS) programme, to develop hardware which would lead to an Earth Resource Operational System (EROS).

In 1969 competitive study contracts were placed with US industry to define and cost the ERTS system requirements, and contracts to build the first two satellites, ERTS A and B (ERTS 1 and ERTS 2 once in orbit) were awarded to US General Electric for launch in 1972 and 1973. The operational system is scheduled to follow within a five-year period, preceded by further ERTS flights.

User requirements

A 'user' is defined as someone responsible for the interpretation or application of Earth resource data. Not surprisingly, requirements are extremely diverse. For example, geologists may wish once and for all coverage of a given area in their work, whereas agriculturalists may want to observe areas of interest several times in the growing and harvesting seasons. Again, foresters seeking a forest fire warning system would not be satisfied with coverage on a weekly or even daily basis, but must receive information on something approaching an hourly basis. The geologists may wish to search for minerals by the coloration the minerals produce in vegetation, but more probably will regard overburden as a screen to observations and hence require a 'defoliating' sensor such as radar. Such a sensor would be unwelcome to the agriculturalist interest in crop studies, although he might welcome radar for land utilization analysis. Detection of forest fires in the latent stages, on the other hand, would call for passive microwave or infrared detectors.

Let us consider first the potential application of remote sensing in geology. Here we find academic geologists, mining geologists, engineering geologists, independent prospecting geologists, mining company exploration geologists and geologists working in teams conducting integrated surveys of developing nations. They have quite different problems. One may seek proof of explanations of continental drift, another oil at great distances below the ground, yet another ore bodies, such as nickel, perhaps only a few feet from the surface. Mining geologists are interested largely in the mechanics of soils to support shafts for mines, or perhaps bridges or motorways. Survey geologists seek to present all they find.

One feature often sought in exploration is major crustal fractures or deep fissures. It is widely believed that these fractures may be the loci of ore concentrations related to deep volcanic activity or simply to higher temperature. These fractures may have no associated offset of geological formations to show their presence, but may be incipient rift fractures, flanked by axes of elongated warps too shallow in dip to be measurable by mapping, or by axes of change in geothermal gradients. Intersection of faults provide the possibility of geothermal energy, or of mineral formation through geothermal plumbing systems. Many ore deposits are still in process of formation, and some are highly exothermic; for example, oxidation of sulphide deposits produces secondary 'hot spots'. Again, the extent of fracturing needs to be known in assessing the reservoir potential of rocks for oil and water.

Widely used techniques for remote survey in geology are gravimetry, seismology, magnetometry and VLF induced transients.

Photo-geology is not as widely used as might be expected, and is always associated with field work. The possibility that *restrahlen* signatures in the thermal infrared can be used to identify rock type is an extremely important one. Infrared imagery has many possible geologic applications including establishing geothermal areas, locating buried dead-ice masses, distinguishing streams fed by meltwater, groundwater and surface water, and studying the location of sporadic bodies of permanently frozen ground near mineralizing ore bodies, and in the surface detection of all these features.

In each discipline the story is repeated. By 'agricultural user' may be meant a Government department taking a census or enforcing a subsidy payment scheme, an agricultural botanist tracing the progress of a crop disease, or the US Narcotics Bureau searching for carefully hidden fields of marijuana or opium poppy. Some users will require pictorial data, while others will be content with numerical data.

It is generally accepted by the layman that interpreters can draw remarkable conclusions from aerial photographs. My experience is that this ability is generally innate with respect to qualitative descriptions. The quantitative use of photographs, however, forms part of the technology of photogrammetry and is a skilled area. Interpreters show little difficulty in moving from aerial photographs to orbital photographs.

Sensor technology

Earlier in this Chapter I outlined the general principles of remote sensing, mentioning several categories of sensor. Any detailed discussion of all sensors and possibilities is ruled out for this book, but I shall indicate the variety of Earth resource sensors available, and offer short descriptions given of two sensors now serious candidates for spaceflight.

Electromagnetic sensors fall into several classes: imagers, radiometers, scatterometers and altimeters. They operate in the ultraviolet, visible, infrared, microwave and radio regions, and may be grouped by technique as cameras, scanners, radars and lasers.

The combinations generally accepted as being of interest for Earth resource work are the metric mapping, conventional aerial, panoramic, ultrahigh resolution and multiband film cameras; the high-resolution television cameras; microwave imagers, radiometers and spectrometers; radar imagers, scatterometers and altimeters; infrared imagers, radiometers and spectrometers; and both visible and infrared laser altimeters and scatterometers. Static force field sensors (gravimeters and magnetometers), on the other hand, although valuable for aerial exploration, lack the discrimination and sensitivity needed at orbital altitudes.

The quality of the television system intended for the first true Earth resources satellites represents a vast technological advance. RCA, for example, has developed a return beam vidicon camera system which combines a storage vidicon photoconductor with an orthicon electron multiplier. The sensor has exceptional lowlight sensitivity, high signal-to-noise ratios and extremely high resolution. It operates with a 4500-line mode, which we may compare with the 405/625 line domestic system and the 800 lines of weather satellite cameras (see Table 7). For an Earth resources mission, three such cameras would be used with 5-inch f/2·8 lenses and filtered to 4750–5750 angstroms, 5800–6800 angstroms and 6900–8300 angstroms, respectively. The system therefore has the information content of near-infrared film.

TABLE 7. Comparison of the television camera tubes in the *Tiros* meteorological satellites and for Earth resources satellite.

	Tiros	*Earth Resources*
Type	*Vidicon*	*Return beam vidicon*
Tube diameter	$\frac{1}{2}$ in	2 in
Lines	400	4000
Line pairs/mm	32	80
Elements	160 000	1 700 000
Sensitivity (ft candle sec)	0·4	0·1

The three-colour elements can be combined to give a single colour or, more precisely, a false colour picture. For permanent film records at the ground – 'hard copy' – it has been necessary to develop image producers. One is the laser beam image producer, also developed by RCA. It uses a helium-neon laser, focused on unexposed film, while its intensity is modulated by the video signals through a ferroelectric crystal. At the same time the laser beam is scanned across the film and the film is driven forward in synchronism, to image a continuous picture.

Fig. 33 shows a multispectral scanner developed by Hughes Aircraft for use in an Earth resources satellite. This sensor uses photomultiplier tubes for three spectral bands between 5000 and 8000 angstroms, silicon photodiodes between 8000 angstroms and 1·1 micrometres, and an intrinsic crystal detector for a band between 10·4 and 12·6 micrometres. The detectivity of the long-wave length detectors is high, but they must be cooled to reduce thermal noise signals in relation to optical (photon) processes.

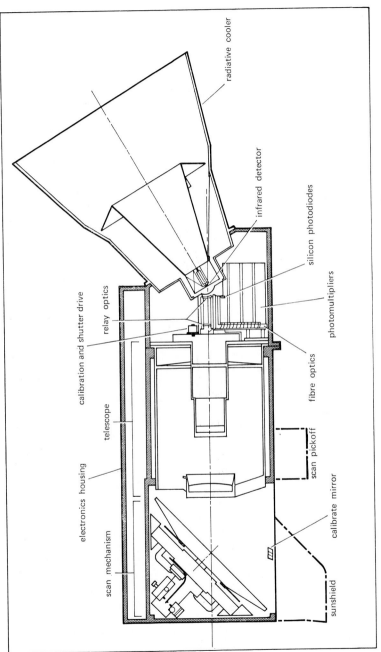

radiative cooler

infrared detector

silicon photodiodes

photomultipliers

relay optics

calibration and shutter drive

telescope

fibre optics

electronics housing

scan pickoff

scan mechanism

calibrate mirror

sunshield

FIG. 33 Multispectral scanner for Earth resources satellites.

Orbits for remote sensing

Three orbital parameters (see Chapter 1) must be considered. First, there are the apogee and perigee heights – hence the altitude above the Earth's surface and the orbital eccentricity. Second, there is the inclination, the angle by which the satellite's orbital plane is inclined to the Earth's equatorial plane. Third, there is the right ascension of the ascending node measured from Vernal Equinox, or the point of the North-bound equator crossed by the satellite, specified by longitude and time.

Mission factors which we must consider are, first, the objectives, that is the terrestrial features we want to observe; and second, the lifetime of the mission. The mission objectives will lead to a definition of the sensors to be carried, so that values can be established for such factors as spectral interval (that is, whether visible band, infrared, thermal infrared or microwave, or more than one band); field of view overall; and spatial resolution.

The sensor definition will then show what is required from the other satellite sub-systems, particularly with regard to attitude control (will the satellite motion about the pitch, yaw and roll axes cause the wrong area to be covered or distortion and blurring in the image?); and data handling (can the satellite tape-recorder store enough data for return to one ground station when the satellite re-appears overhead after several orbits?). Thus we find, for example, that the data down-link from satellite to ground station must be considered as a possible constraint to the altitude.

The lower the satellite, the shorter the time it has to transfer its information for a given bandwidth. Moreover, if certain signals are to be received the satellite must have at least 10 degrees of elevation above the horizon. Of course, we could deploy more ground stations, or use another satellite in a higher orbit, possibly at geosynchronous altitudes, for data relaying.

The field of view of the satellite's sensors delineates a swath across the Earth's surface, and since complete coverage is desirable, a swath-to-swath overlap of about 10 per cent is allowed for. In several discrete altitude regions it is possible to have systematic progression of the daily swaths.

To ensure systematic and controlled data collection – say, for monitoring the growth and ripening of crops – it is clearly desirable that the satellite shall 'see' the ground beneath it under identical conditions of illumination. The effects of cloud cover, dust, aerosols and seasonal variations are, for most regions, secondary to any differences in illumination due to the time of day. It is therefore desirable for passive sensing that the local time beneath the satellite be the same for all areas. Among the advantages are that objects of the same size cast shadows of the same length at the same latitudes, and that any diurnal cycling of crop appearance is eliminated. An orbit in which a satellite appears overhead at a given latitude and longitude at the same time each circuit is termed sun-synchronous. NASA's ERTS A mission requires a circular Sun-synchronous polar orbit at an altitude of 496 nautical miles with a lifetime of at least a year. Orbiting 14 times each day, with the swath seen by the sensor shifting 90 miles westward each day, it will cover the globe every 20 days.

Data handling

The sheer volume of data to be returned from an Earth resources satellite is the central problem. Even from the experimental ERTS, 10^{12} bits per day will be telemetered. This will be presented to the users in the course of one year in the form of 50 000–100 000 colour pictures from the television cameras and in records embracing 1000 million square miles from multispectral imagery, along of course with additional data on point-in-orbit, day, time, attitude and so on. All this, remember, will come from a single satellite with only a few sensors. An operational system might raise these figures by one or even two orders of magnitude.

The strain on communications systems in sending back photographs of high information density point-by-point is immense. In manned Earth resources experiments on *Skylab* 1 in 1973 it is intended to use the astronauts to select areas of interest, allowing for Sun angles, cloud and priorities. The astronaut will operate as a filter, returning only urgent images electronically, the rest returning by shuttle or by separate re-entry capsule.

Fig. 34 illustrates the data management problem that exists at the ground. The essential operations that must be performed on raw data are: de-coding of the telemetered information; normalization and rectification, whereby all data is corrected to the appropriate coordinate system; location in exact geographic terms; information extraction by interpreters; and indexing and storage. Development work is in hand to allow all key operations – including large amounts of the interpretation – to be performed automatically.

Proposed satellites

Most published proposals for Earth resources satellites are adaptations of existing, well-proven designs for meteorological or scientific satellites. Some studies are self-developed, for example *Ceres*, a design based on the capability of the Canadian space industry. This satellite would weigh 250 lb and carry both an infrared linescanner and a

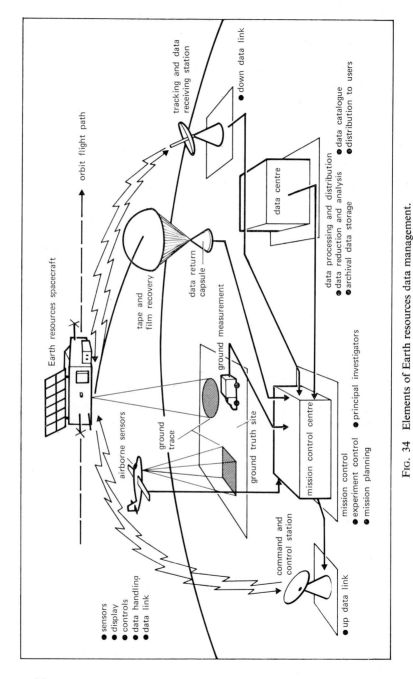

FIG. 34 Elements of Earth resources data management.

high-resolution vidicon camera. The proposed method of attitude control is a modified gravity gradient system. Another satellite and its attendant system would be called *Demeter* (Mother Earth).

A TRW proposal for a low-cost resources survey satellite has been based on the company's *Vela* spacecraft (Chapter 7); and a larger one based on the OGO satellite (Chapter 6). The NASA's ERTS programme is the only one which has reached the point of building an Earth resources satellite.

The ERTS A and B spacecraft, their sensors and the ground data handling scheme are intended to be experimental. They are steps in the development of the operational system to follow. We fully expect, however, that practical use will be made of the greater part of the data they supply. In particular, ERTS A will carry the three-camera return beam vidicon system I have already described, and a four-channel multispectral radiometric scanner. Definition of the data handling system led to the choice of a 20 MHz wide, S-band system and two 10·5 MHz video tape recorders.

The final choice for these first two Earth resources satellites lay between a modified *Nimbus* weather satellite of the US General Electric, and an Earth resources version of TRW's orbiting geophysical observatory (OGO). In the event *Nimbus* was chosen.

I have already mentioned some advantages of having an observer to filter and control information. In addition, he could carry out maintenance and make adjustments, thus increasing the quality of images and the reliability of the system, and reducing costs through increased operational lifetimes. But space station orbits now being proposed reach only about 55 degrees inclination, so it is likely that unmanned satellites of the ERTS type will continue to operate to give coverages at higher latitudes.

Costs and economic benefits

Some debate has already arisen as to whether it would not be cheaper to survey the world entirely by aircraft rather than by satellite. Aircraft are favoured on such grounds as their selectivity and low capital cost, whereas the counter arguments in favour of satellites centre on the great area covered and the value of repeated observations.

There are real difficulties in surveys over the almost constantly clouded portions of the Earth, where an aircraft using film cameras may have to wait for weeks before cloud cover shifts for a few hours at the correct time of day. A full-scale Earth resources satellite system would have a satellite over the area every day at the correct time, and would be able to take advantage of the first cloud clearance. This assertion, of course, begs the question whether the resolutions achieved would be compatible with the survey requirements. Further, in order to detect large-scale features one must build up photographs of large areas from individual aircraft exposures, by constructing mosaics.

Many lineaments have been spotted in satellite photography which would have been most difficult, even impossible, to detect in a mosaic – quite apart from the expense of obtaining compatible aircraft coverage over such large areas.

But let us be clear what is being compared. If the cost per square mile is considered, then the argument in favour of the satellite is 200 : 1 for the first coverage, and the ratio increases for every month in orbit. The costs for each unit of information gathered are extremely difficult to compare, even qualitatively. The true position is that both satellites and aircraft have important parts to play in Earth resources surveys. Introduction of the satellite systems will call for more rather than less aircraft work as the quantity and quality of remote survey increases.

An estimate by the US National Academy of Sciences shows that the total cost for an Earth resources satellite operational system, used in a commonality approach with all disciplines including meteorology would be of the order of £200 million for the non-meteorological portions (assuming about £180 million for meteorological segment). This expenditure would be over a 7-year period and would cover research and development, initial hardware investment and operation, and maintenance in both the space and data distribution and usage sectors. The allocated funds for the ERTS A and B satellites and their associated data management systems are £25·7 million. This is the basis, however, for an operational system and it is possible to consider simpler experimental systems for European purposes which would certainly cost less. To put these sums of money in perspective, world annual expenditure on cartographic surveys alone, by existing techniques, is over £500 million.

Table 8 shows the estimated savings in the first decade of Earth resources satellite operation over existing methods of gathering data

TABLE 8. Estimated Annual Savings, 1975–1980

| | Millions of £ | |
	United States	World
Agriculture and forestry	4 – 16	20 – 24
Geography	2 – 6	4 – 20
Geology and mineral resources	6 – 60	40 – 240
Hydrology and water resources	8 – 20	14 – 40
Oceanography	40 – 160	200 – 360
	£60–£262	£278–£684

in the several disciplines. The economic implications of the Earth resources satellite system have been estimated as shown in Table 9. Costs to the users will depend largely on how much data they wish to handle, and consequently how many interpreters and how much

TABLE 9. Potential annual benefits, 1975–1980

	Millions of £ United States	World
Agriculture and forestry	3×10^2	4×10^3
Geography	4×10	3×10^2
Geology and mineral resources	6×10^2	2×10^3
Hydrology and water resources	4×10^2	3×10^3
Oceanography	2×10^3	3×10^3
	$3 \cdot 3 \times 10^3$	$12 \cdot 3 \times 10^3$

equipment must be employed. The minimum cost of an official national programme of participation might be that of a four-man department; or on the other hand £20 million for full involvement. Data in image form of Earth resources quality from the *Apollo* 9 multispectral photographic experiment is presently available at a cost of approximately one penny (0·42 p) per hundred square miles.

Political and legal aspects

The 1961 resolution of the UN General Assembly recommended the principle that international law, including the UN Charter, applied in space as on Earth, and that the exploration of space be freely open to all nations. But it did not refer to the exploration of the Earth from space, or define the boundary between airspace and outer space. The term airspace is also undefined. The residual atmosphere persists for thousands of miles into space, and the difference between space-flight and the overflying of sovereign states is far from clearcut.

The political issues – apart from the basic rights of nations with regard to surveillance of their territories – are the potential military, strategic and tactical significances of certain information; the unilateral possession and release of economically valuable data; and the use of data for political purposes.

Military reconnaisance equipment is often classified by resolution, instruments with a resolving power below a certain value being free from constraint. More important usually is the area under surveillance – to which access may be forbidden at any resolution. It has been suggested that instruments with a resolution worse than 100 feet provide no data of military significance. This is a highly debatable point, however, and the criterion of significance may be detection rather than resolution. There was considerable disquiet in 1968 about the information content of certain photography from the *Apollo* 7 flight, when reconnaissance-type, high-resolution film was used by the astronauts and, it was reported, matters of national security were recorded.

The impact on world commodity markets of reliable methods of trans-global assessments of crops is very difficult to assess. But as our methods improve, the effect must surely become considerable. Clearly the US Government must be concerned not only that it should use the information acquired by spaceflight in an equitable way, but that it should be *seen* to be doing so. Again, the impact on oil and mineral exploration is also expected to be significant and the nature of this business – where companies normally operate outside their national frontiers – is bound to lead to friction on these issues unless considerable skill is exercised.

On 18 September 1969, President Nixon in an address to the 24th session of the UN General Assembly, said: '. . . we now are developing Earth resource survey satellites, with the first experimental satellite to be launched sometime early in the decade of the Seventies. Present indications are that these satellites should be capable of yielding data which could assist in as widely varied tasks as these: the location of schools of fish in the oceans; the location of mineral deposits on land; the health of agricultural crops.

'I feel it is only right that we should share both the adventures and the benefits of space; and as an example of our plans, we have determined to take actions with regard to Earth resource satellites as this programme proceeds and fulfils its promise.

'The purpose of these actions is that this programme will be dedicated to produce information not only for the United States but also for the world community. We shall be putting several proposals in this respect before the United Nations.'

The future

The Earth resources satellite will come, and many of the claims made for it will be fulfilled. Moreover, I would emphasize that there is no 'Big Brother' element in the satellites I have been describing. Nonetheless, solutions must be found to the political problems of operating the system in the interests of global harmony.

At the present time Europe is considering several possibilities for commitments to participate in several NASA programmes, and it will be interesting to see whether Earth resources surveys figure in this scheme. Europe certainly has the immediate technological ability to build the satellites, sensors and launchers for a purely European programme.

Man has always sought knowledge of his environment. Necessity and curiosity have always caused him effort to identify those features of his surroundings which are helpful, dangerous or interesting. Civilization has demanded increasingly complex and lengthy inventories of these features, and their collection and maintenance has been achieved by laborious field surveys and examinations, often completed too late to be useful. Earth resources satellites within the next decade will give increased knowledge and control of these resources. There will be many benefits, but perhaps the greatest will be humanitarian ones, in that struggling and under-developed nations of our planet will at last have an immediate means of planning the use of their own resources.

In conclusion, millions of human beings have taken thousands of years to build up our present knowledge of the Earth. If Man has any thoughts of exploring and surveying other planets, then it is clear that he must first learn, on Earth, to use the techniques I have described in this chapter.

I would like to thank colleagues Dr John Farrow and Mr Gordon Smith of Hawker Siddeley's Space Division for suggestions, and the Directors of Hawker Siddeley Dynamics for permission to publish.

CHAPTER **5**

Navigational Satellites

R. E. ANDERSON
US General Electric Company

Aircraft fly across the oceans according to previously filed flight plans. Occasionally, a 'blunder' error causes one of them to be far off his planned course. There is no way to detect 'blunder' errors, so the flight plans are arranged to keep a large separation between aircraft. At peak hours, some aircraft must fly routes which are longer than the minimum-time routes in order to maintain the required separations. The added flight time imposes an economic penalty, increasing fuel costs and reducing payload because of heavier fuel load that must be carried. If traffic control could be used on the transoceanic routes, the separations could be reduced while maintaining the present high standards of safety.

Traffic control requires independent surveillance to monitor the relative positions of all the craft in the controlled space, a decision-making authority that knows the positions of all the craft at all times and can decide what action must be taken to avoid conflict, and un-delayed communications between the authority and each of the craft in the controlled space. No means now exists for independent surveillance over the oceans, beyond about 250 miles from the shore, and the presently used high-frequency radio that depends on iono-spheric reflection does not provide undelayed communications.

Ships that pass through congested waters on their approaches to major ports are supposed to sail within the limits of established in-bound and outbound sea lanes. When ships from the open sea enter an outer confluence area several hundred miles from the port, they depend upon their onboard navigation to place them within the limits of the inbound lane. The accuracy of present-day navigation is such that it is possible that an inbound ship may actually be in the outbound lane. A traffic control system for ships could reduce the probability of this dangerous occurrence, and aid in the reduction of the present high incidence of collisions and groundings.

Some military craft need high accuracy navigation on a world-wide basis, but do not wish to reveal their positions. Fishing boats wish to return to a productive fishing spot perhaps only a few hundreds of feet in extent, but they do not wish to reveal the exact location of the fishing spot. Such users require a world-wide, high-accuracy navigation system by which they can determine their own positions. But it must be passive; that is, it must not require transmissions by the user craft. For the fishing boat, it must be inexpensive.

Biologists wish to track migrating animals, perhaps even birds, over large regions of the Earth. Meteorologists wish to track balloons; oceanographers wish to track and read data from buoys all over the world.

Earth satellites may be used to fulfill many of these navigation, tracking and traffic control requirements because they can provide the advantages of line-of-sight radio propagation between distant points on the surface of the Earth, and their positions in space can be determined so accurately that they may be used as position references for precise location systems. Satellites may also be used to relay information between vehicles *en route* and distant ground vehicles. The dual capabilities of position fixing and communications offer the means to implement traffic control over intercontinental distances.

There are four basic position fixing techniques: measurement of distance, rate of change of distance, angle and rate of change of angle. Numerous variations are possible within each technique, and a system may incorporate some combination of the basic techniques. In addition hybrid systems are possible, in which measurements for position determination are obtained by a non-satellite means and the measurements relayed through a satellite to a ground terminal where the fixes are computed. In a hybrid system, the satellite serves only as a communication relay, and knowledge of its position is not required as a reference for locating the craft.

Satellite radio-location systems are also classified as 'active' or 'passive'. In an active system, the user equipment must transmit; in a passive system it does not transmit. A traffic control system must be active, so that the decision-making authority knows the user's current position. An active system is limited in the number of users it can serve, since each must transmit through the satellite. A passive system is useful as a navigation aid by which a user receives signals from the satellites and determines his own position. There is no limit on the number of users that can be served by a passive system. A passive system could be used in traffic control if a separate communication link is provided.

Active systems

In most proposed active systems, an interrogation signal in the form of a user address is transmitted from a ground terminal and is relayed through a satellite. The user's response, also relayed back through satellites, verifies the user's identity and provides information that enables the ground terminal to determine the user's position. The position, which is immediately available for traffic control, may also provide automated navigation if it is transmitted back through the satellite for automatic display aboard the craft.

In another kind of active system, intended primarily for location and data readout of remote sensors rather than for traffic control, the interrogation signal may originate aboard the satellite. The interrogations are scheduled within the satellite by recorded commands received from a ground station. IRLS (Interrogation, Recording, Location System) of the National Aeronautics and Space Administration employs equipment aboard the *Nimbus* satellite to interrogate sensors on remote platforms, such as weather balloons or oceanographic buoys, as the satellite passes over them while circling the Earth in its low orbit. The sensors reply, transmitting the data they have collected since the last interrogation. The satellite measures the ranges to the sensors and records the data. When the satellite makes its next pass over the ground station, it transmits the data and range measurements it has collected from many sensors. The ground station determines the position of each sensor from the range data.

Most proposals for active systems recommend that locations be determined by measuring at least two ranges from the satellites to users (Fig. 35). A range measurement is made by modulating a time marker on a radio signal. The signal is transmitted from a satellite to a user craft, which receives and retransmits it. The time interval is measured from transmission of the marker to the reception of the user's response at the satellite. The time interval measurement is converted to range by multiplying the interval by the velocity of light, and dividing by two. The product is divided by two because the signal travels the distance twice. Round-trip travel time is approximately 12·4 microseconds per nautical mile of range. The measurement must be corrected for time delays in the user's receiver and transmitter, and for a change that may occur in the velocity of radio wave propagation as the signal passes through the atmosphere.

Determination of a user craft location requires at least two range measurements from separated positions in space. Each measurement determines a sphere of position with its centre at the satellite and its radius equal to the measured range. A third sphere of position has its centre at the Earth's centre and a radius equal to the Earth's radius plus the altitude of the craft. The three spheres intersect at two possible locations. They are usually very far apart, so that the correct one can be selected on the basis of prior estimates of the user's location.

The time marker can be introduced on the radio signal in several ways. One way is to transmit a short pulse of radio-frequency energy like a radar pulse. It is difficult to generate high peak power in a satellite, so the transmission of a single short pulse may not be practical. Methods are available for reducing peak power by transmitting

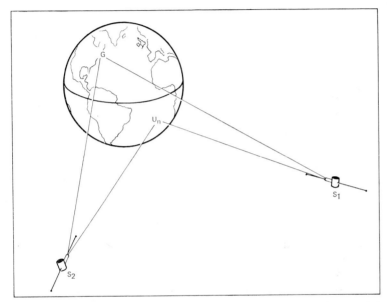

FIG. 35 Range–range active system ground terminal, G. transmits a coded signal to satellite S_1. All user craft receive the signal; the one that is addressed, U_n, responds. Both satellites, S_1 and S_2, relay the user's response. The ground terminal receives the responses, measures radio signal propagation times, determines ranges from known positions of the satellites to user and computes the user's location.

for a longer time while retaining the ability to resolve time with high precision. One method of reducing peak power is by pulse compression. A short pulse is applied to a frequency dispersive element, that is one that delays the frequency components in the pulse by an amount proportional to their frequency so that the high-frequency components are delayed by a different time from the low-frequency components. The result is a pulse that is stretched in time and reduced in amplitude so that the peak transmitter power is reduced in proportion to the time stretching. A reduction of 100 is practical. At the receiver, the pulse is applied to a dispersive element with the inverse characteristic so that the original short pulse with its high timing precision is reactivated.

Another method, now being tested by General Electric in the USA, is 'tone-code' ranging. A short transmission of an audio frequency is used to phase an audio oscillator in the user equipment, which then generates a continuous train of short pulses in synchronism with the zero crossings of the audio tone to provide high-resolution timing. The tone transmission is followed by a digital address code. When the address code is recognized in the user equipment, it allows one of the short pulses to be gated out of the circuit, corresponding to a single radar pulse. After a precisely measured time delay, the user equipment transmits the tone-code signal back to the satellite.

Multiple-tone ranging is another method of reducing peak power. It is used in several applications, including NASA's range and range rate satellite tracking system and the US Army's SECOR (Sequential Correction Of Range) surveying system. Multiple-tone ranging utilizes several audio tones of different frequencies. The tones are transmitted from a ground terminal, through the satellite, to the user craft and return. Each unit receives on one frequency and transmits on another at the same time. The phases of the tones transmitted from the ground station are compared with the tone phases received back from the user craft. The highest audio tone determines the precision of the measurement as propagation time can be resolved to a fraction of the period of one cycle. The lower tones are used to resolve range ambiguity, since the exact number of high-frequency cycles between transmission and reception cannot be known without a coarse range measurement.

Range measurements require at least two satellites for a position fix, but angle measurements can do it with a single satellite. However, it has not been demonstrated that angles can be measured from synchronous satellites with sufficient accuracy to meet most requirements.

A combined angle and range technique, originally suggested by Westinghouse Electric Corporation, employs two interferometers mounted at right-angles on the satellite. An interferometer is constructed by mounting a pair of aerials on the ends of long booms extending from the satellite. The phase difference of the signals received by the separated aerials is a function of angle of arrival relative to the line connecting the aerials.

Fig. 36 relates the phase difference, $\Delta\phi$, at the two aerials, to the direction of arrival. The wavelength, λ, multiplied by the portion of the wavelength represented by $\Delta\phi/360$ yields the difference in distance the wave had to travel from the source to aerial 2 and 1. The difference in distance divided by the length of the baseline is the tangent of the angle of arrival relative to the interferometer axis. A phase difference measurement is repeated for each 360-degree phase change, resulting in an angle measurement ambiguity. The ambiguity is resolved by the use of another interferometer with a shorter base-line, or by a second frequency slightly removed from the first.

Another angle measurement approach has been suggested by

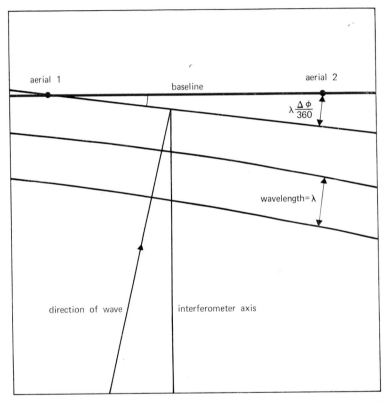

FIG. 36 How phase difference at the two aerials relates to the direction of arrival.

Philco-Ford. Orthogonal fan-shaped beams are transmitted from a rotating satellite. As the satellite turns on its axis, the beams sweep over the Earth in a pattern in which the time between reception of the signals at the user craft depends upon the user's location. The user craft transmits signals back through the satellite if the technique is applied in an active system.

In addition to range and angle measurements, a third active tech-nique is the 'hybrid'. Signals received from existing ground-based electronic aids are transmitted through a satellite to a ground terminal, where fix computations are made. NASA's OPLE (Omega Position Location Equipment) system uses the VLF signals of the Omega hyperbolic system. OPLE has been demonstrated by transmitting Omega signals as they are received at the user's location through an Applications Technology Satellite to NASA's Goddard Space Flight Center. When a satellite is used in a hybrid system, it serves only as a signal relay. It is not used as a position reference; hence, its position does not have to be known accurately.

Passive systems

Passive systems enable the user to determine his own position from signals received with satellites without requiring the user to transmit. It is necessary that the received signals be different at each point in the area served by the system. The user relates a measured characteristic of the received signal to a geographical location, which he then identifies as his position.

Several proposals have been made for one-way passive ranging techniques. Time marker signals are transmitted from the satellite at accurately known times. Each user craft carries an accurate clock, and observes the time of reception of the signal. The interval from the known transmission time to the reception time is determined and converted to range. Measurements from two or more satellites at known locations are necessary to determine a fix, as in the active ranging technique. The one-way ranging technique requires highly accurate clocks on the user craft, such as caesium time standards. An error of five micro-seconds introduces a range error of 1 mile. Clocks that are sufficiently stable to achieve the necessary accuracy for long periods of time are so expensive that the technique is not attractive for most applications.

The requirement for high clock accuracy is reduced if a range difference technique is used. Time marker signals are transmitted from a pair of satellites in an accurately known time relationship. The difference in time of reception is measured at the user craft. The locus of points having the same time difference is a hyperbolic surface of position. The intersection of a hyperbolic surface with the surface of the earth is a line of position, similar to a LORAN line of position. It is much easier and less expensive to measure time differences accurately than to measure actual time accurately, so the time difference technique is usually preferred to one-way ranging. Two separated satellites are required for a single line of position, so that at least three satellites are required for a position fix.

Angle measurement techniques can be adapted to passive systems. A 'radio sextant' employs an aerial with a high-resolution beam pattern that measures the angle between local direction references and a radio source. If the receiver is sufficiently sensitive, it can be used with the Sun, Moon or radio stars as well as satellites. It is useful only at microwave frequencies where the high angular resolution can be achieved with a small aerial.

The rotating fan beam satellite could be operated in a passive mode. The user would measure the time difference as the fan beams sweep by him, and compute his own position. Similarly, transmitting interferometers on satellites can be designed to have lobe patterns that sweep over the Earth and provide a pattern of received signal variations that are related to position.

A passive approach in operational use is the US Navy's Transit system. High-accuracy fixes may be obtained several times a day at every point on the Earth. While it serves its intended purpose very well, it is not useful for air traffic control because an aircraft can move quite far in the time the satellite is observed to gather data for a fix, and the velocity of the aircraft affects the Doppler measurement. Fixes are not always available with the present constellation of satellites, and a separate communication channel would be necessary for traffic control.

Transit uses low-orbit satellites and the Doppler technique. The satellite transmits signals at 150 and 400 MHz that are frequency-stabilized to one part in 10^{10}. It also transmits information in digital form about its orbit so that the user can determine the satellite's location as a function of time. The user observes the change in received frequency due to Doppler's principle as the satellite passes above his horizon. Fig. 37 depicts the received frequency during a pass of the satellite. The largest change of frequency occurs when the satellite passes directly overhead. It is smaller for passes to one side or the other of the user's position, so that the rate of change of frequency can be used to determine the user's distance from the sub-satellite track. The received frequency is the same as the transmitted frequency and there is an inflection in the curve of frequency versus time at the closest approach.

In principle, the user may first determine the track of the sub-satellite point from ephemeris data received from the satellite, and then determine his lateral displacement from the track by a rate of frequency change, and his position along track by the time of the inflection of the frequency shift curve. Rotation of the Earth modifies the relationship of received signal characteristic and location in a way that resolves the ambiguity between the two sides of the sub-satellite track. It is not necessary to receive the signal during the entire satellite pass, nor to receive during the inflection to obtain sufficient information for a fix.

Simultaneous reception of the two radio frequencies, 150 and 400

MHz, is used to determine corrections for ionospheric propagation delays when users require position fix accuracy within a few hundred feet.

The US Navy is experimenting with Timation, a system refined beyond Transit, that incorporates one-way range measurements as well as Doppler measurements. It is also capable of distributing

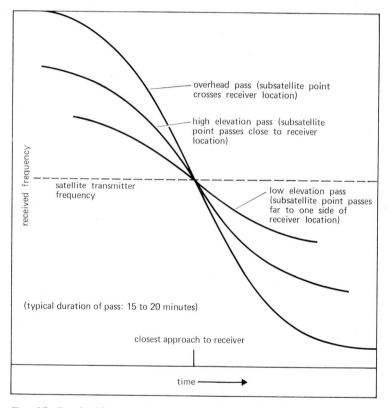

FIG. 37 Received frequency during a pass of the satellite. The greatest change in frequency occurs when it is directly overhead.

timing signals with an accuracy of the order of microseconds to all parts of the world.

Active ranging and position fixing techniques have been tested using NASA's Applications Technology Satellites ATS-1 and ATS-3. These satellites are in geostationary orbit, 19 300 nautical miles above the Earth's equator. ATS-3 is above the Atlantic. Its position over the equator is moved occasionally between approximately 75° and 45° west longitude. ATS-1 is above the Pacific, at approximately

150° west longitude. Voice communication to aircraft and ships has been demonstrated extensively through the satellites.

Side tone ranging from the satellites to the German research ship *Meteor* has been successfully tested by the West German Institut für Satellitenelektronik.

An extensive series of tests has been performed by the US General Electric Company using the tone-code ranging technique. The experiments demonstrated for the first time that it is practical to locate mobile craft by range measurements from two satellites with a single interrogation from a ground station and a single response from the mobile craft. A short tone-code interrogation signal containing a single frequency tone burst followed by a user's address was transmitted from a ground terminal to ATS-1 or ATS-3 (usually the latter). The satellite repeated the signal. All of the activated transponders within range of the satellite received the interrogation and each one matched the phase of a locally generated tone to the received tone phase. The one unit that was addressed recognized its own code, and transmitted a response through an omni-directional aerial. Both satellites repeated the response when both were within view of the unit. The ground terminal measured the time from the initial transmission to the return of the interrogating signal from the one satellite, and to the returns from the user as relayed by the two satellites. The time measurements were stored on punched tape and later inserted into a computer. The known equipment delays of the user transponder and the satellites were subtracted from the measurements, and the ranges from the two satellites to the user were determined. An initial fix determination was made, the local time at the initial fix was noted and corrections for ionospheric delay were obtained from a model of the ionosphere stored in the computer. Range corrections were applied, and the position fix determined by another iteration of the computer.

When a transponder was in view of only one satellite, lines of position were determined from the single range measurements. Computer programs were used to compute the latitude at which the line of position crossed a given longitude, or the longitude at which the line of position crossed a given latitude.

Other data were also collected during the experiment, such as the standard deviations of range measurements between the ground terminal and the satellite, and between the satellite and transponders aboard mobile vehicles or at fixed locations. Ionospheric delays were measured, and observations were made of Faraday rotation, ionospheric scintillation and sea reflection multipath. Voice and data transmissions were made with the same transmitters and receivers used for ranging.

Five vehicles were used in the tests: two aircraft, a DC-6B and a

KC-135 of the US Federal Aviation Administration; a US Coast Guard cutter in the Gulf of Mexico; a buoy moored in deep water off Bermuda; and a panel truck in upstate New York. Each was equipped with conventional mobile communications transmitters, receivers and antennas and had a 6-in by 8-in by 10-in, 6-lb experimental tone-code 'responder' unit attached between the receiver and transmitter. The combined receiver, transmitter and responder is termed a 'transponder'.

When a vehicle was to be located, a ground station transmitted a 0·43 second tone-code signal to one of the satellites, the 'interrogating satellite', usually ATS-3. The signal consisted of a 2·4414 kHz tone burst followed by the individual user address formed by suppressing an audio cycle for a digital 'zero' and transmitting an audio cycle for a digital 'one'. The tone-code signal was frequency modulated on a 149·22 MHz carrier with a narrow deviation so that the r.f. bandwidth was within the 15 kHz bandwidth of the mobile receivers.

The satellite repeated the signal on 135·6 MHz. All of the activated vehicle equipments received the signal, and each matched the phase of a locally generated audio tone to the received tone phase. The one vehicle that was addressed responded with a short burst of its properly phased, locally generated tone followed by its address code, introducing a very precisely known time delay between reception and retransmission of the code. The vehicle response on 149·22 MHz was through a broad-beamwidth aerial. If both satellites were in range of the vehicle, they both repeated it on 135·6 MHz.

The ground station received the returns from the two satellites separately with narrow-beamwidth aerials. It measured the time interval from its initial transmission of the signal to the first return from the interrogating satellite, and the two returns from the satellites as they were relayed back from the user. From these measurements the ranges from the two known positions of the satellites to the vehicle were determined. These ranges, together with vehicle altitude and corrections for ionospheric delay were used to compute the vehicle location. When only one satellite was in range, a line of position was computed and a fix defined as the crossing of the line with latitude or longitude of the vehicle determined by other means.

The time required for the interrogation and response was less than one second except when a data transmission was included with the user response or, for the aircraft, the equipment required a longer time to switch from receive to transmit. The usual interrogation rate was once every three seconds, although a once-per-second rate was demonstrated.

The tone-code ranging technique has the following characteristics:
● Useful accuracy can be achieved within the modulation and r.f. bandwidths of present-day mobile communications.

- The technique can be used with wide bandwidth for high accuracy.
- It requires only one channel for range measurement, receiving and transmitting in the simplex mode if desired without need for an aerial diplexer.
- The time required for a range measurement is a fraction of a second, so that it can time-share a communication channel with little additional time usage of the channel.
- It can be implemented by the addition of an inexpensive, solid-state responder unit attached to a communication receiver-transmitter.
- It can, but need not, employ digital or digitized voice transmissions to provide synchronizing of the user responder, thereby further increasing the efficiency of the channel usage.
- There are no 'lane' ambiguities in the range measurements.

User identification is simple and is confirmed in the return signal. The test results indicate that an accuracy of ± 1 nautical mile, 1 sigma,* for ships and aircraft can be achieved at VHF. To achieve that accuracy, it would be necessary to employ calibration transponders at fixed, known locations with approximately 600 mile spacing, and interrogate each one a few times per hour to determine range measurement corrections. It is recommended that calibration of vehicle equipment time delay be accomplished at the ground terminal by interrogating each craft when it is at some known location. The time delay calibration is then stored in the computer with the vehicle address. It would be necessary to employ aircraft aerials that discriminate against sea reflections, so that the reflected signal is more than 10 dB below the direct signal. The use of circular polarization for the satellite and aircraft aerials is recommended.

The VHF frequencies, 135 and 149 MHz, used in the ATS-1 and ATS-3 experiments are near the lowest that can be considered for position fixing from satellites. They are economically attractive because suitable aircraft communication equipment is available for use with satellites. For example, the new Boeing 747 large jet aircraft are furnished with aerials and communications designed for voice links through satellites at frequencies between 127 and 136 MHz.

Ionospheric propagation effects and signal bandwidth limitations imposed by frequency assignments limit the accuracy that is achievable. VHF mobile radio-frequency bands are crowded. The aircraft band, 118-136 MHz, is fully occupied with channels at 50-kHz spacings. It has been suggested that present-day frequency control

* Sigma states the confidence level of a measurement. If the errors have a normal or bell-shaped distribution, 67 per cent of the fixes will be within one nautical mile of the time position if the accuracy is 1 mile, 1 sigma.

permits the use of 25-kHz channel separations, and that satellite communications can be assigned some of the interleaved channels that would then become available. Specific proposals for such assignments are now under consideration.

L-band frequencies, 1540 to 1660 MHz, have been assigned, but are not currently in use for operational air-ground communications. They offer relief for the present crowding of the VHF band, and for satellite location systems they offer the potential of higher position fix accuracy. For these reasons, L-band will be required for future use. The cost of implementing a satellite system at L-band and the cost of user equipment will be higher than at VHF because more development is needed. Aerials which are physically large enough for L-band use have narrow beamwidths, thus requiring that they be pointed toward the satellites. This complicates the design and increases the cost, especially for aircraft.

Several systems have been proposed for L-band, and experiments suggested to test the system concepts and evaluate signal propagation.

The French are developing the Dioscures system, a worldwide interrogation and radio determination system. It is a two-satellite ranging system that uses L-band for transmissions between satellites and user craft. The range measurements are made by transmitting a continuous digital code to the craft. All aircraft receive the code and synchronize clocks aboard the craft. When a craft is interrogated it retransmits the code in phase with the received code. The Earth stations measure the time between transmission and reception. The transmissions include multiplexed digitized voice and digital data.

NASA's PLACE (Position Location and Aircraft Communication Equipment) system uses multiple tone ranging for locating user craft by range measurements from geostationary satellites. Voice and digital communications can be combined with the ranging signals.

The SPOT (Speed POsition and Track) system of RCA is a multiple tone ranging system similar to PLACE. It is intended for applications at L-band frequencies. Binary phase shift keying is employed on one of the tones to serve as a digital data channel. RCA has suggested the use of one-way ranging with accurate user clocks as a passive mode for the SPOT system.

NAVSTAR and other systems studied by TRW can be used in active and passive modes by ranging or range differencing methods. An interesting satellite constellation is suggested for use with the systems. If a satellite were placed in an inclined elliptical orbit with a 24-hour period, with proper selection of the orbit inclination and ellipticity its ground track will trace a circle with its centre at a point on the Earth's equator. It has been suggested that three satellites be placed in orbit so that their ground tracks trace a circle, with the satellites phased in the orbit so that the ground tracks are spaced along the circumference of the circle. A fourth satellite is placed in a geostationary orbit so that its subsatellite point is at the centre of the circle. The satellites in the constellation are then deployed so that range and range difference measurements from the satellites can provide accurate location, velocity, altitude and rate of climb for the user craft.

As evidenced by the many techniques described above, technology is available for implementation of worldwide navigation and traffic control systems to meet a variety of user requirements. There are already many position location needs that could be fulfilled by the use of satellites and for some applications such as transoceanic air traffic control and confluence area ship traffic control, satellites appear to be the only technically and economically feasible solution.

The Earth implementation of satellite systems for mobile terminal communications and position location is not delayed by technical considerations, but by the need for agreements by many parties on the functions that the system will provide, what agencies will be responsible for the operation of the system, and the frequency bands in which it shall operate. The radio-frequency spectrum is already crowded and the added burden of satellite communication and position location for large numbers of mobile terminals requires the careful consideration of international agencies which are responsible for allocating the use of the spectrum. These problems are under continuous serious consideration by national and international deliberative bodies.

Preparatory work on frequency allocations by many countries has been coordinated through CCIR, the International Radio Consultative Committee, in preparation for the World Administrative Radio Conference to be held in 1971. Many studies are under way to anticipate future requirements for satellite systems. Satellite technology offers such a wide range of useful applications that matching user requirements and technical possibilities for maximum economic advantage is a major challenge.

CHAPTER **6**

Research Satellites

DR P. STUBBS
New Scientist

Space research to date has been primarily concerned with the exploration of the near-Earth environment and the solar system. Not surprisingly the largest number of satellites has been devoted to investigating those phenomena nearest to the Earth – those of the upper atmosphere and ionosphere, and of the surrounding belts of trapped radiation and the magnetosphere (Fig. 38) of which they

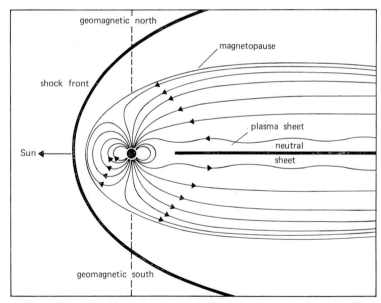

FIG. 38 A major achievement of space research has been to define the Earth's magnetosphere, the vast magnetic envelope which surrounds us. On its sunlit side a plasma shock front repels the perpetual wind of solar protons; and a long tail stretches out on the night side of the Earth.

form an intrinsic part. Because the behaviour of these entities rests heavily upon that of the Sun and its electromagnetic and corpuscular radiations, these, too, have received considerable attention from the space scientists.

As regards our immediate neighbours, again, we have expended more time and effort exploring the Moon than the nearer planets and, as yet, none on the more distant planets. However, with the first decade of space research well past, there is a growing tendency to employ the advantages provided by working beyond the confines of the Earth to study the cosmos far more widely. True space astronomy, which previously has largely concerned itself with solar physics, is now beginning to blossom. And current plans and recent achievements such as the spectacular Soviet success in remote sampling of the Moon's rocks promise exciting new possibilities in planetary investigations.

Let us, to start with, take a closer look at just what the space environment does provide in the way of opportunities for pure research which are peculiar to it. There are four broad areas for scientists to exploit here.

Upper atmosphere measurements

Perhaps the most obvious is the use of satellites for direct on-the-spot measurements of the Earth's upper atmosphere and of interplanetary space. Before the 'space age' opened in the autumn of 1957 man could obtain knowledge of these regions only by inference. Although he had exploited the radio-reflecting layers of the ionosphere for long-distance transmission for several decades, measurements of important properties of these layers, such as their electron density and temperature, were only possible by indirect means such as radiosondes. The lowest part of the ionosphere begins at a height of some 40 to 50 miles, well beyond the reach of the highest sounding balloons. Likewise ideas about the types of ion that made up the layers were largely guess work.

About interplanetary space ignorance was even greater. Before the early satellites we knew nothing about the Van Allen belts of trapped energetic particles surrounding the Earth (*Explorer* 1 is credited with this discovery in the beginning of 1958); we had no concept of the solar wind tipping its steady stream of corpuscular radiation out into space and on to the Earth; nor yet of the interactions of this plasma with the Earth's magnetic field, generating the so-called magnetosphere with its characteristic shock front facing the Sun, and long drawn out 'tail' downwind.

Today space geophysicists have an increasingly cogent picture of all these phenomena and their dynamical aspects – how they vary in space and time, in extent and composition, and what effects solar events have upon them. Other facets of this enormous advantage of being able to 'get up among' the happenings in near space include that of being able to measure magnetic fields *in situ* – an essential component in all studies of gas plasmas – that of getting directly at the cosmic-ray primary particles arriving from space (beneath the Earth's atmosphere we can only study the secondaries resulting from showers triggered off in it by the primaries); that of probing the thickness and nature of interplanetary dust which, too, does not survive penetration of the atmosphere; and that of being able to measure the upper atmosphere's density from drag effects.

Continuous monitoring

One of the greatest facilities provided by satellites is that of continuous monitoring on a global scale. This aspect is of importance especially in the fields of meteorology and Earth resources (Chapters 3 and 4). But it is equally rewarding in ionospheric and magnetospheric studies where few events happen in isolation and where a researcher wants comprehensive data on how, say, the ionosphere reacts to the changing solar radiation as the Earth rotates, what atmospheric tides look like, or how a solar flare perturbs the magnetosphere. Only by employing orbiting satellites can we get the necessary coverage to see these phenomena in full perspective.

A satellite's orbit is fixed with reference to the stars but the Earth rotates beneath it. Thus, in a sufficient number of orbits, its trajectory weaves a basket around the entire Earth. Depending on the apogee and perigee chosen, the shape of this basket can be designed so that it fulfils special purposes.

Extraterrestrial astronomy

The third great advantage of satellite techniques is that of being able to make observations from a station above the Earth's atmosphere. Unmanned space observatories are already opening up extensive new vistas for the astronomer which are quite comparable to that initiated by the advances of radio astronomy since the Second World War. The reason is by now becoming rather well known. The Earth's atmosphere is opaque to all but a small fraction of the wavelengths making up the electromagnetic spectrum.

It is transparent, of course, to visible light having wavelengths between 3900 and 7800 angstroms, and this provides the reason why living creatures have evolved organs sensitive to these colours, and these alone. The atmosphere also provides another window in the radio frequencies between wavelengths of about 16 cm and a few kilometres. Apart from these fairly small 'holes' the atmosphere lets through some microwaves partially and certain bands in the infrared part of the spectrum – although astronomy in the latter generally requires dry conditions and high observing stations.

All other kinds of electromagnetic waves are absorbed before they reach us. But this does not mean that stars, galaxies and other cosmic

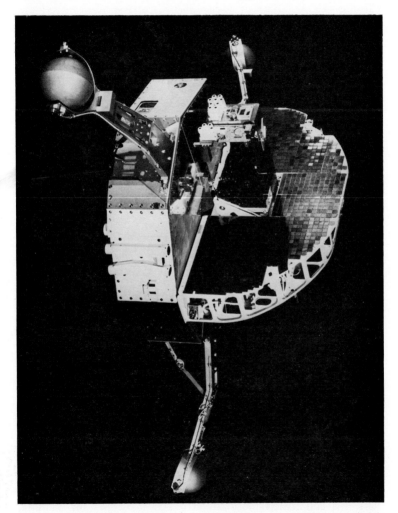

FIG. 39   *Orbiting Solar Observatory* 5, launched in January 1968, carried eight experiments to study the Sun. One of six highly successful satellites, it measured solar radiations which do not otherwise reach the Earth. *OSO* 4 returned the first complete ultraviolet map of the Sun. (*Courtesy, NASA.*)

I

bodies do not emit them. There is thus an as-yet largely untapped wealth of information about the universe awaiting the extraterrestrial astronomer. So far, it has been exploited to a considerable extent by rocket-flown equipment – responsible for all but the most recent work on X-ray stars – and by two series of successful satellites. These are the six *Orbiting Solar Observatories* (Fig. 39) which have mapped the Sun for the first time at X-ray and ultraviolet frequencies; and *Orbiting Astronomical Observatory* 2 (Fig. 40) which, since December 1968, has measured the ultraviolet brightnesses of many thousands

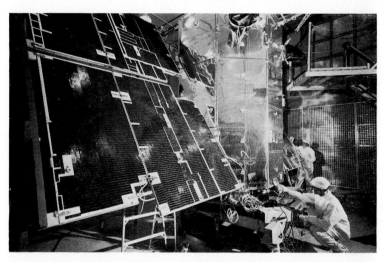

FIG. 40 The two-ton, 21-foot-wide *Orbiting Astronomical Observatory* 2 launched in November 1968, carried 11 telescopes designed to make the first ultraviolet survey of the whole sky. (*Courtesy, NASA.*)

of stars in an overall sky survey, and examined in detail some hundreds of ultraviolet objects. A few inconclusive gamma-ray experiments have been tried and two successful radio-astronomy satellites. As yet infrared satellites for astronomical purposes have made little headway, though they are employed in planetary science and for meteorological and military purposes.

In principle, of course, space can provide the astronomer with near-perfect 'seeing' conditions well above the perturbations of the atmosphere and away from interfering light sources. Until space platforms can match those of terrestrial telescopes for stability, however, pointing accuracy will be inadequate to exploit this advantage. On *OAO* 2, for example, experiments can only be pointed to within about half a minute of arc – relatively crude compared to the precision of the best modern ground-based telescopes.

Escape from gravity

The aspect of an orbital environment which has greatest appeal to the public imagination is undoubtedly that of weightlessness or, in other words, free fall. For experimental purposes its advantage lies in the ability to remove one factor – gravity – which is inescapable within the confines of a terrestrial laboratory. In the experiments of physics this is no great drawback in general; you can predict the role of gravity with certainty in nearly all cases. But in biology and medicine it is by no means always sure where it exerts a controlling influence. To what extent are reproducing cells or germinating seeds affected by the presence of a gravitational field? How is plant growth dependent on gravity? How does the addition of space radiation affect the result?

These are some of the questions space biologists have tried to answer, particularly Soviet researchers and their American counterparts responsible for the *Biosatellite* series of spacecraft. In living animals and man similar queries arise over the control of the cardiovascular system and the vestibular functioning that determines orientation.

Free fall, however, is of use for two aspects of physics at least. Precise tracking of spacecraft in orbit has added immeasurably to the geophysicist's knowledge of the Earth's shape (Fig. 41) and the global distribution of mass, for every satellite in orbit responds to every minor irregularity of the gravitational field. This data is invaluable for theories of the internal structure of the Earth. And the principle can be extended – and in the case of the Moon has been extended – to vehicles in orbit about other bodies.

The other application is more specialized. By measuring the very small annual precession which a weightless and frictionless gyro makes in free fall it ought to be possible to check one of the conclusions of Einstein's General Theory of Relativity. Development work is at present under way at Stanford University to design a satellite experiment of this kind.

Planetary spectaculars

I have now listed advantages of the space environment which might be summarized as: getting up into things, getting right round things, getting above things and getting away from things. There is one other – getting out to things – the spectacular field of planetary probes, lunar spacecraft and other deep-space vehicles. This expensive research has called forth great ingenuity and led to some of the most impressive results of the whole of space research. The modern generation of planetary explorers has crash-landed and soft-landed

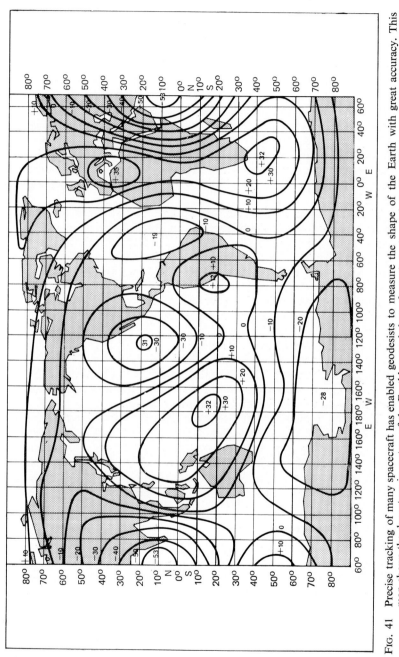

FIG. 41 Precise tracking of many spacecraft has enabled geodesists to measure the shape of the Earth with great accuracy. This map shows the departure in metres of the Earth's actual shape from that of a perfect ellipsoid.

vehicles on the Moon and Venus, carried out reconnaissance fly-bys of Venus and Mars, put distant spacecraft into orbit around the Moon and, latterly, drilled out lunar samples and returned them to Earth – all by remote control. Man is unlikely to set foot on any planet other than Mars and that not in the near future. These

FIG. 42 *Lunar Orbiters* are responsible for extensive detailed photographic mapping of the Moon and determinations of its precise shape. A *Mariner* Mars orbiter was launched in 1971. (*Courtesy, Boeing Company.*)

113

various automatic techniques for coming to closer grips with our solar system neighbours, however, have already supplied much new data and seem likely to yield much more yet, given enough money of course.

Among the techniques available for unmanned planetary exploration, photography – or more correctly the transmission of vidicon television pictures – has played the largest single part to date. Much of the material selenologists continue to debate, even after the manned lunar landings, arises from the superb lunar close-ups col-

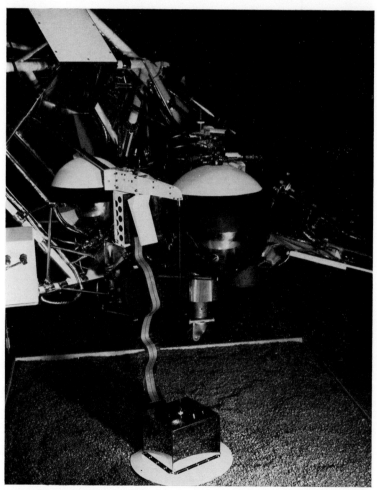

FIG. 43 (a) and (b)   The ingenious alpha-particle scattering device aboard *Surveyor* 5, which supplied the first analysis of lunar soil composition.

lected by the five *Lunar Orbiter* spacecraft (Fig. 42). Our only reliable knowledge of the appearance of the surface of Mars comes from the pictures that *Mariners* 6 and 7 sent back in 1969 (and to a lesser extent *Mariner* 4 in 1965). In addition these *Mariners* carried infrared radiometers that successfully measured the Martian polar-cap and other temperatures, and determined rates of cooling; and infrared and ultraviolet spectrometers to study the compositions of the planet's atmosphere and surface – both successfully accomplished. A radio occultation experiment in which S-band waves were transmitted through the atmosphere on the limb of Mars measured its radius and confirmed the previously observed flattening, in addition to finding the density of electrons in its atmosphere.

On Venus the most advanced research has been performed by the Russians who, with *Veneras* 4, 5 and 6, dropped equipment by parachute through the Venusian atmosphere, determining its composition, temperature and pressure. Antedating the *Apollo* venture *Surveyor* spacecraft performed a number of highly ingenious, if simple, experiments to test the lunar surface, in addition to adding substantially to the Soviet score of lunar surface close-up pictures obtained with *Luna* 9 in 1966. The five out of seven successful *Surveyors* dug 18-inch trenches in the lunar soil, bounced on it with force transducers attached to their legs, searched for lunar magnetic particles, gave us the first hints of the soil colour by photographing

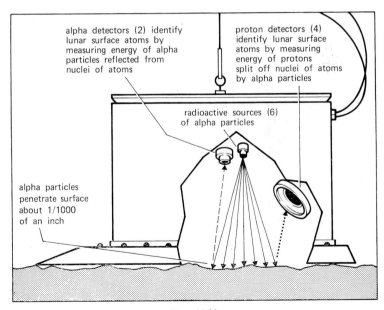

alpha detectors (2) identify lunar surface atoms by measuring energy of alpha particles reflected from nuclei of atoms

proton detectors (4) identify lunar surface atoms by measuring energy of protons split off nuclei of atoms by alpha particles

radioactive sources (6) of alpha particles

alpha particles penetrate surface about 1/1000 of an inch

FIG. 43(b).

115

it through filters, and made the first determination of its composition with the aid of a cunning alpha-particle scattering experiment (Fig. 43(a) and 43(b). There seems little to prevent these techniques being used on Martian soft-landers eventually. Two such vehicles are planned for 1973 under Project *Viking*. A Mars *Orbiter* should reach the planet in November, 1971. One additional bonus to come from planetary exploration with space probes is the much more accurate determination of the astronomical unit, the Earth–Sun distance which establishes the precise scale of the solar system. Relative distances are known simply from celestial dynamics to a very high degree of accuracy, but one good dimensional determination is necessary to give the absolute values.

Some techniques of space research

So much then for the kinds of study which you can undertake in space. I will now go on to discuss some of the techniques that are used with satellites and the snags which arise, describing a selection of vehicles by way of examples. In a chapter of this length it is plainly impossible to be comprehensive but there are plenty of more detailed accounts available. All I shall attempt here is to indicate some of the more salient features, and the kinds of dodges it may be necessary to get up to in order to make experiments work.

Basically much of the actual scientific equipment that goes into satellites is simple. It has to be light in weight, of course, and robust enough to stand the mechanical stresses of a rocket launching and the environmental extremes of space. These requirements may call for special materials and design, but most of the devices operate on familiar laboratory principles, except that many are developed to work only in a restricted range of values pertinent to a specific objective – if you are looking only for water vapour you only need a spectrometer capable of measuring the appropriate absorption bands. Measurements in space are mostly concerned with some kind of particulate or electromagnetic radiation and therefore it is hardly surprising that a large proportion of the apparatus flown on satellites consists of energetic particle counters, electrostatic fluxmeters and ionization gauges; or spectrometers, radiometers and photo-detectors. Various kinds of shielding and filters provide discrimination for the wavelength or particle energy range desired in a particular experiment. To detect interplanetary dust earlier experimenters used electrical conductivity measurements of thin plates which were gradually eroded away by the micrometeoroid impacts. Since then they have tried piezo-electric detectors, microphones and, in the case of the very large *Pegasus* experiment, extensive arrays of capacitors.

Often the simplest kinds of measurement are the hardest to perform. Radio-astronomy satellites have been notoriously difficult to design because of interference picked up from the rest of the spacecraft equipment by the highly sensitive aerials. The UK satellite *Ariel* 3 was the first to overcome this limitation successfully, to be followed shortly afterwards by the US *Radio Astronomy Explorer* 1 (*Explorer* 38). This latter satellite is of particular interest for its extreme development of an aerial fabrication technique that has proved essential for long dipoles in space. Preformed, hard alloy tape, some two thousandths of an inch thick, is deployed in space from a motor-driven reel so that it coils up into a long thin tube perhaps half an inch in diameter. On *RAE* 1 the two longest of the three dipole aerials so formed were each 1500 ft long. Even more extensive schemes have been mooted, though. One was for a spider's web-like radio dish, 12 km across, to be erected in space.

Likewise magnetic measurements always pose particular problems. Magnetic fields in space are very weak in general, and sensitive electronic 'fluxgate' or rubidium-vapour instruments are employed. But stray electric currents from accompanying equipment on the satellite can easily play havoc with their readings if they are not carefully positioned, usually out on the end of a boom.

Thus the design of research satellites, usually carrying a number of experiments, is in large part a matter of integrating the different functions and doing so within the specified payload limitations. In addition, there are such factors as the thermal balance of the spacecraft in relation to the positioning of the equipment; and the siting of equipment to fit in with the chosen spin mode of the vehicle. Then there is the question of data transmission and retrieval, and how each experimenter is to dovetail his requirements into those of his colleagues in the most economical manner. It is easy to see that, just taking these few factors into account, the planning of a satellite is an arduous business.

Attitudes and temperatures

A satellite's spin can serve in the electronic sense to 'chop' signals arriving at detectors for purposes of amplifying them. But its main function is to stabilize the vehicle's attitude, and attitude control is one of the most vital aspects of all classes of research satellites. It is hard to think of any likely experiments that would not call for a knowledge of the direction in which the equipment was pointing.

Attitude control needs the highest precision for astronomical observations. More complex, however, are the requirements of planetary probes. Unlike a simple Earth orbital satellite, which can be orientated in a fixed direction so that its solar cells receive enough

117

sunlight for power, a planetary probe like the *Mariner* craft requires a total of six Sun sensors, feeding information to the active attitude-controlling gas jets to keep its solar paddles aligned in the most advantageous position. It must also have an Earth sensor to keep its high-gain receiving and transmitting aerial pointing towards the Earth. Mariner has three different planetary sensors to cater for different phases of the Mars or Venus encounter to keep its cameras and other equipment trained on the planet correctly. And, throughout a long trajectory, the vehicle navigates by a lock-and-follow type of star tracker, usually employing the bright star Canopus.

OAO 2, by contrast, obtains its accurate attitude by six gimballed star trackers mounted on three control axes. These feed error signals into inertial wheels on each of the axes, which then accelerate and rotate the spacecraft by reaction into the desired position.

Passive spacecraft orientation is also possible in principle by gravitational means, using a long boom with a weight on the end. The vehicle is then sensitive to the gravity gradient and the weight should keep it 'upright' like a fishing float. In practice this technique has not met with a great deal of success, but another – using the field lines of the Earth's magnetic field – is more feasible and will be employed by the British satellite *UK* 4, scheduled for launch in 1971. Its forerunner, *Ariel* 3, established a novel system for ground monitoring of its attitude. It consisted simply of a set of six special mirrors which were used in conjunction with the light-reflecting solar cells disposed on the spacecraft body and four booms. The pattern of sunlight glints from these various reflectors enabled ground observers to fix the attitude of the spin axis to within less than 2 degrees.

Thermal control in an environment in which external heating and cooling take place entirely by radiation poses a constant problem to the designer of a satellite. Always the side looking towards the Sun tends to overheat while the other tends to cool too much. Both 'active' methods involving controlled louvres, and 'passive' techniques with insulation and paint, can be introduced to overcome this extreme situation. The 1969 *Mariners* are described as maintaining temperature control, 'by combination of thermal louvres, deployable sunshade, aluminized Teflon insulation, paint patterns and polished metal surfaces'. Thermally, they are clearly no simple design.

Choosing the orbit

The selection of orbits for most experimental satellites is more crucial than for applications satellites (see Chapter 1), which in general demand either a circular or geostationary orbit. For probing near-Earth phenomena highly elliptical orbits may be desirable if you require, say, to slice a section right through the magnetosphere, as in the case of the European Space Research Organization's *HEOS* 1, which had a high-inclination orbit with apogee at 13453 miles and perigee 2696 miles. Alternatively, a circular orbit will suit you best if you want to study latitude variations at one altitude. Other considerations may involve the length of time for which the satellite is in Sun and shade.

Latterly, experimenters have come to consider another factor – the possibility of using satellites in combination – and this imposes further conditions on the choice of orbit. It has proved to be a very powerful way of studying the near-Earth phenomena because simultaneous measurements in different parts of the magnetosphere, the ionosphere or the radiation belts can give a much more complete picture of an event.

One classical example concerns a solar storm that occurred on 11 January 1967. Five satellites – *Explorer* 33, *ATS* 1, *Velas* 5 and 6 and *OGO* 3 – all distributed in various orbits about the Earth, made observations of the effects of this major solar flare eruption upon the solar wind and the magnetosphere. They revealed a compression of the magnetosphere which was followed by a rebound to abnormally large dimensions. Likewise the Canadian satellite *Alouette* 2 was launched with *Explorer* 31 in 1965 to make complementary measurements of ionospheric composition and temperature variations. (Canadian satellites, including the *Alouette* series, have been remarkably successful in probing the upper layers of the ionosphere from above by making use of the radiosonde techniques normally performed on the underside from the ground.) Ionosphere physicists are also interested in making pairs of simultaneous observations at conjugate points, the 'opposite ends' of terrestrial magnetic lines of force.

At the other extreme is the need to make simultaneous measurements of related factors but all in the same place. Particle densities, for instance, may have a bearing on radio phenomena; or the optical manifestations of the aurora on magnetic field strength. And there has been a tendency more recently to design satellite payloads to meet these specific requirements. In the most extreme case this philosophy leads to the concept of the omnibus satellite; one which contains a large number of experiments, mostly on related aspects of space, which achieve a measure of economy from possessing common facilities such as power sources and communications and telemetry equipment.

The best example is undoubtedly provided by the US *Orbiting Geophysical Observatories* (OGO). The six vehicles of this class which NASA orbited between 1964 and 1969 carried a total of 130 experiments connected with the solar wind, solar flares, geomagnetic disturbances, sudden ionospheric disturbances, radiation-belt particle populations, auroral events, ionization and atmospheric density

119

variations. Between them these box-like, ungainly half-ton satellites made many notable discoveries. The *Orbiting Solar Observatories* (OSO) and *Orbiting Astronomical Observatories* (OAO) are built upon the same principle – a standard spacecraft into which can be slotted the desired experiments. *OSO* and *OAO*, however, carry considerably fewer experiments than the *OGOs*. In rather similar manner, the later *Interplanetary Monitoring Platforms*, such as *IMP G* launched into a highly elliptical orbit in June 1969, carried a whole range of equipment. *IMP G* set off to explore the magnetosphere with 12 experiments aboard.

In conclusion, it seems that the first round of space exploration is now over. No startlingly new phenomena are likely to be discovered in near space, though there are plainly some to come from both the inner and outer planets as well as the Moon. The coming decades, however, are certain to see a great refinement in techniques and an enormous increase in our knowledge of the workings of inter-planetary space.

CHAPTER **7**

Military Uses of Satellites

NEVILLE BROWN
University of Birmingham

A few years ago Dr Harold Brown, then the Secretary of the US Air
Force, argued that 'We must gear our space activities to functions
that can be more economically based in space and to activities that
cannot be performed in the classical manner.'* It is probable that
today there would be very general agreement among military specialists to this effect. Yet not so long ago men were to be found in several
countries who saw space vehicles not just as useful auxiliaries to land,
air and sea forces but as the harbingers of a fourth medium of warfare.

Speaking in January 1958, Brigadier-General Homer A. Boushey,
the US Air Force's first Director of Advanced Technology, cited six
reasons for his acceptance of the contention that 'He who controls
the Moon, controls the Earth'. The reduced gravitational pull of the
Moon would, he argued, facilitate missile launches. Surprise attack
from the Earth on lunar installations would be impossible. An
absence of blast would be only one of several advantages to be
derived from the lack of a lunar atmosphere. Outposts on the far side
of the Moon would always be invisible from the Earth.†

Nor was this remarkable presentation by any means the only
triumph of enthusiasm over analysis in respect of the military implications of the advent of the space age. Four years later one British
writer went so far as to envisage the fate of civilizations being
decided by chivalrous tournaments in space.‡ Meanwhile, the first
post-war Soviet official textbook on strategy laid stress, particularly
in its second edition, on the need for such space-based or space-

---

* *General Electric Forum*, Autumn, 1967.
† Speech reproduced in Eugene Emme, *The Impact of Air Power*, Van Nostrand,
1959, Chapter XII, p. 865.
‡ M. Golovine, *Conflict in Space*, Temple Press, London, and St Martin's
Press, New York, 1962, p. 117. See also *Journal of the Royal United Service
Institution*, November 1962.

oriented systems as orbital bombs and anti-satellite devices. Some of the worst Western journalistic science fiction was alluded to in its discussion of anti-gravity techniques, lasers and the like.* About the same time some Soviet jurists and diplomats were contending that any State had the right to shoot down any space satellites manifestly being used for 'espionage' against it.†

Although this sort of technological 'hawkishness' may now be extinct some significant vestiges did survive until recently. One of the several very distinguished British specialists on Soviet strategy has suggested that talk about a 'military revolution' in 1968 may well be related to the launching of the *Soyuz* 4 and 5 docking space vehicles early in 1969.‡ Concurrently Curtis Le May, the first commander of Strategic Air Command, was upbraiding his political leaders for neglecting, among other things, the need for military superiority in space:‖ one clear indication, in his view, of this neglect being the cancellation of the *Dyna Soar*, the single-seat experimental machine that was intended to orbit in space and then land by means of aerodynamic surfaces and controls.

But isolated counter-attacks along these lines have not prevented the widespread acceptance – not only in the United States but also elsewhere – of the more cautious approach epitomized in the statement by Dr Brown cited at the start of this chapter. Undoubtedly one important factor in this crystallization of opinion has been the emphasis placed since 1960 in Washington on comparative cost-effectiveness as a key criterion for the allocation of defence resources. Another has been the development of better political communications, and what generally appears to be a more relaxed relationship, between the USA and the USSR. A more specific consideration has been the movement towards a negotiated system of arms control, as exemplified by the Strategic Arms Limitation Talks and also by the treaty prohibiting the placing of nuclear weapons in outer space and the use of the Moon and other celestial bodies for military purposes. This treaty was endorsed by the UN General Assembly in December 1966 and signed by Britain, the USA and the USSR the following month.

Space surveillance

But to note the current disinclination to regard military activities in space as anything more than auxiliary to those elsewhere cannot be to dismiss them as unimportant. For in one particular field – surveillance – the space era is already having fundamental effects. The

* V. D. Sokolovsky, *Military Strategy*, Moscow, 1963, p. 394.
† Quoted by R. K. Woetzel, *International Relations*, April 1963.
‡ John Erickson, *The Listener*, 20 February 1969.
‖ Curtis Le May, *America is in Danger*, Funk and Wagnall, 1968, p. 277.

Americans have used unmanned satellites to reconnoitre from space since 1961 and the Russians since 1963. Most of the vehicles employed revolve for one or two weeks in something like polar orbits at heights of between 60 and 250 miles; they collect their data either by optical or infrared photography or else through the interception of radio signals. Although the US Air Force never reveals how many reconnaissance satellites it launches per year, the total may well run into dozens; the USSR is believed to have launched 135 by the middle of 1969. Some of the US Air Force vehicles are thought to be able to manœuvre in orbit.* Since this facility is important as a means of ensuring ready examination of limited areas, it is quite possible that the corresponding Russian craft have already acquired it too.

Often the Americans have low-definition results transmitted to Earth immediately by means of television; they then get their high-definition results through the ejection of capsules of film which, though weighing several hundred pounds apiece, are recaptured by specially equipped aircraft as they descend through the lower atmosphere. The Russians, too, use television-relay for meteorological and such-like purposes. For military reconnaissance, however, they seem to depend on a controlled descent of the whole vehicle at the end of its active life.

Typically a reconnaissance satellite revolves round the Earth about fifteen times a day; and, during the daylight phases, whatever optical cameras it has can photograph strips perhaps 70 miles wide. But what degree of detail is obtained in this way? Little has been said officially and in public about this crucial question, but enough information that is both authoritative and reasonably precise has become available in other ways to permit us to make a reasonable assessment.

When good-quality film is used in conjunction with high-grade optical lenses, an image definition of better than 100 lines per millimetre can be obtained. Some of the largest airborne cameras in operational use have focal lengths of around 5 feet but certain models have been constructed – at least for experimental purposes – by means of which considerably better results are obtained by the use of prisms and reflecting mirrors. A camera with a definition of 200 lines per millimetre and an effective focal length of 10 feet would, in principle, be capable of distinguishing a light patch from a dark background if the patch was about 9 inches across and the camera stationary 85 miles above: that is, a ground resolution of 9 inches from that altitude. Apparently it is difficult, however, to move the film in the camera in such a way as fully to offset the fact that a reconnaissance vehicle in something approaching a polar orbit may have a velocity of tens of thousands of feet per second relative to any

* *Jane's All the World's Aircraft: 1969–70*, Sampson, Low, Marston, 1969, p. 653–4.

point on the Earth's surface. Therefore, the general assumption is that most American reconnaissance satellites still do not achieve an optical ground resolution of more than 1 to 2 feet. Soviet performance in this regard is probably closely comparable.

What a resolution of this order means in terms of the detection and identification of militarily significant objects is not easy to gauge. It is as well to remember, however, that the Americans were able to map Soviet intercontinental ballistic missiles (ICBMs) accurately as soon as they began routine observations of this kind; and ever since then, of course, the relevant technologies have continually been improved. Today it is often intimated that satellite cameras can usually spot aircraft on runways and vehicles on roads and, in many cases, provide specific information about them. Perhaps, indeed, this is about the level of target discrimination one might expect from optical resolutions of around 1 foot and with the concurrent use of 'false-colour' and other modern methods of photographic discrimination (see Chapter 4).

Infrared surveillance

Not that optical surveillance is the only technique available in the satellite reconnaissance programmes. Automatic radio receivers have proved able to monitor quite low-power transmissions made on short-wave bands at ground level. Then again a great deal of work has been done, at least in the United States, on the use of infrared photography for certain specific tasks. Early in 1960 the first launch of a 4000-lb satellite was attempted in the Missile Defense Alarm System programme (MIDAS), which was intended to detect ascending Soviet ICBMs within the first minute or so after their blast-off by picking up the characteristic electromagnetic signature of their rocket exhausts. Development of MIDAS was halted early in 1963, the conclusion having been drawn that too little was known about certain aspects of infrared spectroscopy to ensure positive identification of an ascending ICBM. But interest in the basic concept has remained alive.

Infrared sensors are also able to record heat from factories or other industrial sources or to detect local changes in the surface temperature of the sea of the order of 0·5°C. Changes of this order may be caused by the passage of a submarine at high speed and within about 200 feet of the surface. Then again, detectors sensitive to the shorter-wave electromagnetic emissions – X- and gamma-rays – and others sensitive to neutron efflux can be used to monitor nuclear tests in the atmosphere or outer space. Both types of cell are fitted in the *Vela* and *Advanced Vela* nuclear detection satellites operated by the United States since October 1963 (that is, shortly after the signing of the Partial Test Ban). The *Vela* model, which was withdrawn from service in 1970, was 56 inches wide and weighed 510 lb; its 90-lb payload consisted of 18 separate detectors which enabled it, in principle, to register nuclear explosions as far away as Venus. It moved in a near-circular orbit of roughly 67000 miles radius. The *Advanced Vela*, which was first launched in 1967, weighs 730 lb including the 155-lb payload. Presumably it can provide more data about the character of atmospheric nuclear tests still being conducted by China and France. About six *Advanced Velas* in 60000 mile orbits are needed to give a continuous coverage of the Earth's surface.

The first two series of American reconnaissance satellites were known as *Discoverer* and *Samos* respectively. The first *Discoverer* was successfully orbited in February 1959 and *Samos* just under two years later. Both were designed by Lockheed and so, in fact, was MIDAS.

New nomenclature

The United States has now replaced the above nomenclature by a series of numbers. Although little has been said about this officially, it is generally understood that there is a 770 class which is designed for wide-angle optical reconnaissance, and a 920 class with narrow-angle cameras (that is, with limited coverage but correspondingly high definition). Meanwhile, the vehicles in the *Vela* programme have reportedly been designated the 823 class, and those used for the continuing MIDAS-type experiments 239. Two years ago the 949 Integrated Satellite commenced operations; 'integrated' meaning in this context that it incorporates the full range of automatic surveillance techniques.

Around the same time another new class – the 817 – began operating in geostationary orbit near S.E. Asia. By a geostationary orbit is meant one that is approximately circular in the plane of the equator and at an average of about 20000 miles above it. It can readily be shown that under those circumstances the velocity and direction of the satellite's rotation keeps its position constant in relation to any point on the Earth's surface (Chapter 1). Apparently it has been found in surveillance that the extra precision obtained from the virtual absence of relative motion more than offsets that lost as a result of the much greater range involved. Most probably the reason for this is that telescopic lenses can be employed comparatively easily if the field below is effectively static.

Russian satellites

Virtually all the unmanned satellites the USSR has sent into Earth orbit have been launched by the military and the great majority go by the generic name of *Cosmos*. But neither the Russians nor the Americans officially release much information about them. Therefore, the exact configuration of the military reconnaissance programme can only be a matter of conjecture. Even so, it does seem probable that *Cosmos* satellites in 51-degree, 65-degree and 72-degree orbits are used for military reconnaissance. Most of them are believed to have weighed about 7000 lb and to have been recovered after about seven days, though since 1968 they have also operated a satellite that normally stays up for twelve days.* As is the case in other major branches of military technology the Russians are unlikely to have so large a variety of systems in operation as does the United States. As has already been implied, however, there is no reason to believe that the photographs obtained are much inferior in quality to the majority of their American counterparts.

Space 'platforms'

Clearly the Superpowers stand only at the threshold of surveillance from space, when one thinks of the possibilities inherent in large manned platforms, perhaps in geostationary orbit and with big downward-directed telescopes on board. Admittedly tactical movements would often still be hidden for a while by such factors as cloud and darkness and by the problems of 'read-out'. But within a few days at the most it ought normally to be possible to detect any development of strategic or economic importance in even the most closed society. Perhaps, indeed, the new 10-ton Low Altitude Surveillance Platform the US Air Force was originally due to orbit 100 miles up late in 1970 represents a major step in this direction.

In order to understand better some of the implications of this prospect it may be instructive to consider the extraordinary impact reconaissance from space has had, and continues to have, on the strategic balance. Thus in June 1962 Robert McNamara, then the US Secretary of Defense, indicated in a speech at the University of Michigan that for some time ahead the United States could reasonably hope to defeat the USSR in any major conflict by means of attacks concentrated on 'all of the enemy's vital nuclear capabilities', namely the launching pads of her ICBMs. His confidence in this new strategy of 'total counterforce' stemmed primarily from the way in which, using her new reconnaissance satellites, the USA had pinpointed all the Soviet Union's ICBMs and had shown them to be few in number, too large and complicated to fire at short notice, and lacking adequate ferro-concrete hardening against nuclear blast. In other words, the Soviet ICBM echelon was vulnerable to virtually

* *Jane's All the World's Aircraft: 1969–70*, Sampson, Low, Marston, 1969, p. 655.

complete destruction on the ground in the event of a strategic rocket attack of the scale and precision the United States had become capable of launching.

The USSR's bomber bases were similarly vulnerable. Meanwhile, her ballistic missile submarine fleet was still at a very early stage in development. So it would seem that the Soviet attempt to move medium range ballistic missiles (MRBMs) into Cuba that autumn should be seen not as a piece of a bold adventurism by a Superpower who felt secure beneath her nuclear umbrella but rather as a panic measure designed to redress a serious adverse shift she perceived taking place in the central arms balance, largely as a result of her adversary's introduction of space reconnaissance. Correspondingly their weakness in intercontinental terms was almost certainly a key factor in the failure of the Russians to press ahead with this MRBM movement in the face of an American challenge.*

But by about 1966 the USA had lost this counterforce option against the USSR because the latter's strategic forces had become a great deal larger and more sophisticated. Today the Soviet ICBM echelon includes over 1000 missiles, many of them in well-hardened emplacements and ready to fire at short notice. Therefore it matches its American counterpart in numbers and quality. Likewise her new but steadily growing flotilla of Y-Class ballistic missile submarines are closely comparable to the Polaris boats.

Much discussion now takes place about just how stable this state of virtual strategic parity between Moscow and Washington in fact is; and in several respects this debate still hinges upon an assessment of the role of space reconnaissance. The facts that Soviet ICBMs normally take about a year to install, and that during this time their sites can be observed, must help to reassure the small but rather articulate minority of Western commentators who still fear the USSR will suddenly attain a significant superiority in this field. Then again the ability of anti-ballistic missiles (ABMs) to disturb the current stalemate is circumscribed by reconnaissance satellites. For the more that is known about the character and distribution of ABM batteries, and the associated master radars, the easier it is to prepare suitable penetration tactics and devices. Deterrence through the threat of retaliation is thereby preserved.

Submarine surveillance

Sometimes it is suggested that infrared detection of the passage of nuclear-driven submarines moving at high speed and within about 200 feet of the surface undermines the role of any ballistic missile

* See the author's 'Towards the Superpower Deadlock', *World Today*, September 1966.

submarine force as a stable deterrent. Yet only a little thought is required to realize how unlikely it is that this technique will ever be able to monitor satisfactorily a sizeable force of boats well dispersed across the oceans. What if the vessel is moving only slowly and at a depth of, say, 1500 feet? What about cloud cover? Suppose the submarine glides beneath an ice-cap? What if a heavy sea is running? How well can these sensors discriminate against reflected sunlight and other spurious signals? What can be done about positive identification? How long does data of this sort take to transmit and interpret?*

But these are not the only ways in which space reconnaissance might be said critically to influence world stability. Thus, although the USA is no longer able to deliver a total counterforce blow against the USSR, both these countries can be expected to retain that option against China for a decade or so to come – a prospect from which some alarming inferences can be drawn. Another disconcerting reality is that the constant collection by 'spies-in-space' of masses of general intelligence data seems bound to cause offence, given the almost obsessional concern so many nations have for their national privacy. What should the United States, in particular, do with all the data it is amassing? To release select portions of it would be to expose itself to charges of news manipulation. To disseminate it beyond the official domain on a confidential basis would be to appear guilty of a new form of patronage. Yet to make it freely available would be to arouse still more widespread antagonism. The peculiar way Washington responded to the first Israeli charges of Egyptian violations of the 'standstill' agreement in the Canal Zone in August 1970 can be taken as some guide to the dilemmas that ultimately may arise.

Satellite interception

Obviously it is not inconceivable that exasperation with being thus spied upon could induce certain governments to try and destroy the offending space vehicles. Indeed, it does seem as if a limited number of the *Cosmos* launches carried out in the course of the last few years have been concerned with the use of satellites to inspect and perhaps destroy other satellites. Thus *Cosmos* 249 and 252 – two manœuvrable spacecraft which went into broadly similar orbits on 20 October and 1 November 1968 – appear to have blown up whilst making a pass at a target satellite orbiting some 300 miles above the Earth. Likewise for some 10 years the United States has maintained, on a low budget and a concept-evaluation basis, a Satellite Interception

* See the author's chapter in Ken Twitchett (Ed.), *International Security: Reflections on Survival and Stability*, Oxford University Press, 1971.

(SAINT) programme, chiefly in order to be ready to reply in kind should an American spacecraft be attacked.* It is well recognized, of course, that it is very difficult to close to contact with an object travelling at over 25 000 feet per second and perhaps taking evasive action; it is also appreciated that to close at a controlled relative speed of perhaps 10 feet per second and to stay within, say, 50 feet sufficiently long to conduct an inspection would be a great deal more difficult. Nor is the acquisition problem even as simple as this. Anti-radar paints exist which are able to reduce the radar cross-section of militarily significant satellites to about 1 square centimetre. An alternative stratagem would be to place dozens of decoys in orbit.

For some years past there has been general agreement that the support elements of any satellite interception should be surface-rather than space-based. By the same token we are unlikely now to witness any revival of interest in the ballistic missile boost interceptor (BAMBI) concept – one of several highly exotic notions in the field of strategic deterrence explored by the United States Air Force in the early 1960s. In essence, BAMBI represented an attempt to solve the ABM problem by placing something between 800 and 3600 satellites in permanent low orbits so as to be ready automatically to launch homing weapons against ICBMs during their several minutes of powered flight. But study of BAMBI was discontinued in 1963. With both ABM and SAINT research now firmly anchored to the Earth's surface, it is tempting to think of systems being developed which are dual-purpose, that is both anti-missile and anti-satellite. However, a fundamental objection to this approach is that nuclear warheads would probably be deemed an undesirable element in any satellite interception system, whereas they are almost indispensable for ABMs.

Orbital bombs

What then of orbital bombs? No such weapon would be worthwhile for anybody unless it had a nuclear warhead. Therefore, true orbital bombs are precluded by the 1966 UN Treaty. But fractional orbital weapons are not, so the USSR began testing a Fractional Orbital Ballistic System at the end of 1966 and brought it into operational service about 18 months later. By 1970 the large SS-9 intercontinental rocket was being used for FOBS test-firings.

A ballistic trajectory is, almost by definition, the most economical for an intercontinental missile. However, the warhead climbs to a height of well over 500 miles and so is normally visible on enemy radar screens throughout the second half of its 30-minute flight. The point about a FOBS, on the other hand, is that it is given enough

* It is not clear whether the code-name SAINT is, in fact, still in official usage.

acceleration to pass into a low earth-orbiting trajectory, for example 75 to 125 miles above the surface. Then it is brought down on to target by the burning of a retrorocket. A FOBS cannot be expected to have more than a minor fraction of the warload of a standard ICBM. Nor is recourse to a retrorocket likely to render it anything like as accurate. Nor, indeed, is it easy to see how, under present circumstances, an ability to clip intercontinental warning times to below ten minutes would confer much tangible advantage on the Russians. What they may have in mind, however, is a pre-emptive strike against the bomber bases of the US Strategic Air Command.

A true orbital bomb complete with retrorocket would share the payload and accuracy limitations of FOBS; it would also pose uniquely serious problems of command and control. For this reason, as well as on account of the respect presumably being accorded the 1966 treaty, weapons of this kind are unlikely to be in service today. All the same, it would only be sensible to bear in mind the risk that they will make their debut in the wake of further nuclear proliferation. Let us consider, for example, the hypothetical case of an emerging nuclear power which lacks an area or position appropriate for the creation of a land-based deterrent, and which is also without an industrial base quite broad or sophisticated enough to build ballistic missile submarines. She might find her answer was an orbital deterrent.

Satellite services

Various other military applications of satellites discussed in earlier chapters merit special mention. Thus geodetic satellites played a crucial part in completing the exact measurements of the Earth's surface that have proved essential to the standard accuracies of less than half a mile now being registered by, for example, ICBMs. Meteorological satellites now flash their observations to Saigon to facilitate air operations over Vietnam. Navigation satellites enable ballistic missile submarines to fix their locations to about a tenth of a mile – at any rate in environments in which it is tolerably safe for the submarine to protrude an aerial above the surface. The US Air Force is now trying to apply them to aircraft navigation as well.

But more important than these three categories, at least from the standpoint of future development, is the role satellites have started to play in both strategic and tactical communications (see, for example, Fig. 44). Thus in July 1967 the US Army declared its Initial Defense Communication Satellite Program to be operational for the Pacific area; and by the end of 1968 this global network – the key element of which is 17 nearly synchronous satellites – was providing five 'voice-equivalent' channels 24 hours a day to terminals in

FIG. 44 Experimental communications satellite designed for the US Department of Defense. This powerful 1600-lb satellite can handle the equivalent of 10 000 two-way telephone channels. (*Courtesy, Hughes Aircraft Company.*)

Hawaii, the Philippines, Vietnam, Germany, Ethiopia, California and New Jersey. Meanwhile, a triservice tactical satellite communications was under development, the aim being to have more consistent contact with such small and mobile stations as those in aircraft and jeeps.

Moreover, transmissions are essential to carry the large volume of radio communications a modern battlefield involves; they also have to be used to transmit the television signals that are now helping to disperse a little of Clausewitz's 'fog of war'. Yet until the advent of the military satellite these transmissions seemed subject to the crippling limitation of being unable, as a rule, to travel beyond the optical horizon. Now 'high-quality' photographs can be transmitted from Saigon to Washington in a matter of minutes by means of satellites. Correspondingly global transmissions in the high-frequency band are becoming more consistently efficient and reliable, and also less dependent upon intermediate stations, as a result of the advent of satellites. Yet as the Chinese have reminded us by their establishment of a satellite monitoring station in the strategically-placed island of Hainan, messages sent via military communications satellites are quite susceptible to interception and also to jamming. Military historians may wonder, in any case, whether it is so excellent an idea to have full and instant communications between the commanders in the field and their superiors in remote rear areas; more than one battle has been lost through the interference by the latter this permits. What must be acknowledged, on the other hand, is that since modern limited conflict is so dominated by political and global constraints a large amount of such 'interference' is almost inevitable. And the better informed it is, the more constructive it will tend to be.

Manned orbiting laboratory

To what extent will exploitation of the varied military potentialities of space involve the further entry of Man into that environment? In 1963 Mr Robert McNamara announced work had begun on what was termed a Manned Orbiting Laboratory (MOL); two years later President Johnson revealed McDonnell-Douglas were to design and develop this spacecraft for the US Air Force. The plan was to provide a laboratory module able to afford a two-man crew a 'shirt-sleeve environment' for up to thirty days and link to it a modified version of *Gemini* 2, a smaller spacecraft which might be used to transport this crew backwards and forwards from Earth. In 1967 the MOL programme was still being officially described as the '. . . best opportunity to obtain knowledge of Man's ability to function in space and to determine the application of that ability to defence purposes', to quote Dr Harold Brown.

Even by that time, however, the first MOL launching had been postponed beyond the following year. Then in June 1969 what would have been about a $3000 million programme was cancelled 'for economy reasons'.

A salient weakness of MOL was terms of reference which were too ambiguous. Wheareas some senior officers in the US Air Force had been thinking of manœuvrable command posts in space* Mr McNamara and his civilian advisers almost certainly had more limited aims. Besides, MOL was tending to be overtaken by advances in the relevant automative techniques. Nevertheless, a safe assumption would be that human beings will gradually come to perform more tasks in Space, especially in connection with the burgeoning art of space reconnaissance. Maintenance is one area in which human aptitudes may be so applied, and the monitoring of observations is another.

* For example, General T. S. Powers, *Air Force and Space Digest*, June 1963.

CHAPTER 8

Advanced Satellite Concepts

Dr P. E. Glaser
Arthur D. Little Inc.

The early history of space flight has seen the realization of one of mankind's greatest aspirations. The first glimmer of Man's chance to convert his fanciful notions of extraterrestrial flight into an idea with engineering significance came with the demonstration of rocket propulsion. The departure of Man and his machines into space and the successful landing on the Moon have already had a profound influence on human thought and the general view of Man's place in the universe: Man has again proved that he has not reached the limits of his creativity.

Once Man has completed the exploratory phase in space, the quickening pace of technological change will spur him on to use space technology to benefit mankind on Earth. Already, as previous chapters show, satellites are beginning to be integrated into Earth-based technology, making their contributions in areas which touch the lives of most of the Earth's population. Today, however, we are only at the threshold of their uses and applications.

For the first time, for example, we can visualize methods to harness solar energy on a very large scale and thus reach out to tap the tremendous potential this inexhaustible energy source can provide for mankind. Utilization of the Sun's energy is an old dream. We have always depended on the Sun but its benefits have been confined to its historical natural function of making the Earth hospitable to life; for example, by maintaining temperatures within a suitable range, by making plants grow and by making rains fall. The enormous energy flux which the Sun provides has long invited collection and conversion into other useful forms, such as mechanical and electric power. While past attempts at such conversion have succeeded on a small scale, they have not shown enough economic promise to find widespread application. Now, however, when people in the industrially developed countries, as a result of a steadily deteriorating environment caused by the technological conveniences of life such as the internal combustion engine and power generating plants, are becoming increasingly concerned with the quality of their lives, the generation of power from solar energy could have a major impact.

TABLE 10. Solar collector surface area required to supply annual electrical energy demand* at 10 per cent conversion efficiency†

	(Square miles)		
	1966	1980	2000
World	1050	2040	6300–9460
United States	330	860	2840
Northeast	60	160	380
South Atlantic	60	140	340
New York City	10	20	60

* See Table 11.
† Solar energy available in orbit assumed to be 0·14 W/cm².
Source: Arthur D. Little, Inc., estimates.

TABLE 11. Annual consumption of electrical energy (10^{12} kWh)

	1966				1980				2000
	T	F	N	H	T	F	N	H	T
World	3·34[a]	—	—	—	6·5	—	—	—	20–30
United States	1·20	0·95	0·05	0·2	2·70	1·80	0·6	0·3	9·0
Northeastern United States	0·22	—	—	—	0·50	—	—	—	1·20
South Atlantic United States	0·20	—	—	—	0·45	—	—	—	1·08
New York City	0·04	—	—	—	0·08	—	—	—	0·19

T – Total; *F* – Fossil Fuels; *N* – Nuclear Power; *H* – Hydro-electric Power
[a] – 1965

Sources: Data from *Statistical Abstract of the United Nations*, US Dept. of Commerce and *Statistical Yearbook*, United Nations.
Starr, C., and Smith, C., *Energy of the World of 2000 A.D.*, presented at National Academy of Engineering, Sept. 1967.
Arthur D. Little, Inc., estimates.

Let us first look at the total solar collector surface area (Table 10) which the countries of the world would require to supply the electrical energy demand projected for the year 2000 (Table 11). If we assume a 10 per cent conversion efficiency, the solar collector area for the United States would be about 2840 square miles – not so large that the technical feasibility would be *a priori* questionable. (In fact, the power plants and associated transmission lines to be constructed by 1990 will require at least this land area.) However, with collectors located on Earth the large-scale generation of power from solar energy appears to be impractical because even areas where solar energy is received throughout most of the year would require either energy storage systems more effective by orders of magnitude than batteries, spinning standby power generators or pumped storage to fill in for interruptions of solar energy.

Since Earth-based receiving stations do not appear to be practical, it is worthwhile to consider the technical and economic feasibility of satellite solar power stations, since they would avoid most, if not all, the undesirable environmental conditions created on the Earth by other methods of power generation.

System technology for satellite power

Although satellites for solar energy conversion such as those shown in Fig. 45 may be several decades away, to guide future developments we can examine the feasibility of specific system design concepts and explore several aspects of the required technology:

- A satellite capable of being placed in an orbit which permits the solar energy receiving area to be exposed to the Sun most of the time.
- Lightweight solar-energy conversion devices with high efficiencies.
- Transmitters capable of transferring the converted energy to an Earth station in a spectral region where minimum atmospheric absorption and scattering would be encountered.
- Earth stations capable of accepting the required power density and transmitting the energy to power distribution networks.

Systems engineering usually is devoted to determining the size or capacity of each component of the system and to predicting the performance of the assembled systems under various anticipated operating conditions. To carry out the required operations requires that the system be well-defined and that its components be identified as to function and type, the sequence in which they are connected well known, their functions analysed and differences between various types reduced.

However, for a satellite solar power station system (Figs. 46 and 47) only part of this information can now be provided, mainly because the technology for several major components changes so rapidly, in contrast to other types of systems (for example, power plants, ships and aircraft) where the technology usually remains constant during the engineering phases. Therefore, technological forecasts have to be made before the engineering phase begins. Although candidates for most of the components now exist, at least in the form of laboratory models, new and quite different components are imminent. How each will perform if fully developed is uncertain,

Fig. 45 Artist's impression of a satellite solar power station.

but systems engineering techniques can be used to approximate and compare the performance and costs of the various system configurations, to estimate the dependency of the performance on its components' characteristics, and to set quantitative targets for component development based on forecast of component technology and performance.

137

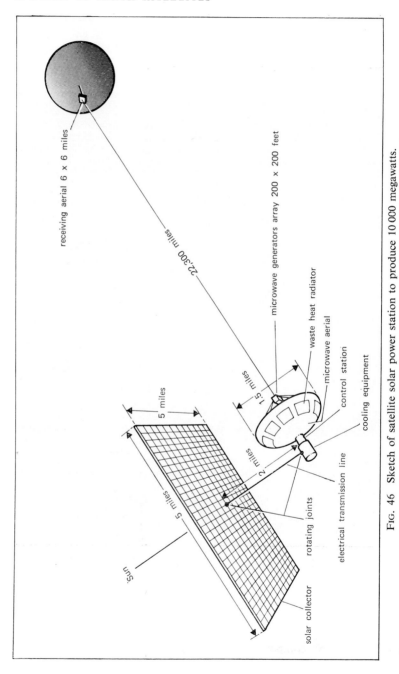

FIG. 46 Sketch of satellite solar power station to produce 10 000 megawatts.

receiving aerial 6 × 6 miles

22,300 miles

microwave generators array 200 × 200 feet

waste heat radiator

microwave aerial

control station

cooling equipment

1.5 miles

5 miles

5 miles

2 miles

rotating joints

electrical transmission line

Sun

solar collector

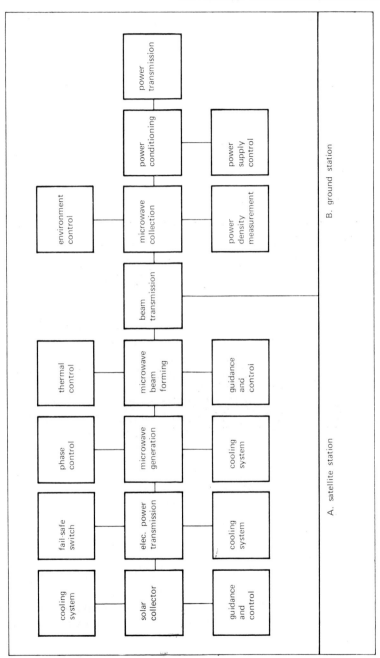

FIG. 47   Major components of a satellite solar power station system.

Because transporting the equipment into orbit and performing maintenance on it there will be considerably more expensive than manufacturing it, its size, weight and efficiency will be of primary interest. The size and weight of the ground equipment will be comparatively unimportant, but efficiency will again be of primary interest – optimization procedures specify that the minimum investment for the satellite station is inversely proportional to the ground equipment efficiency. Therefore, the primary objectives of any satellite solar station will be minimum weight per kilowatt for the satellite components and maximum efficiency for the Earth component.

Conversion of solar energy

To date, photovoltaic solar cells have been the primary source of electric power for almost all unmanned spacecraft, both for space exploration programmes and the application of space technology to communications, navigation and meteorology. Furthermore, in recent years, improvement of the technology of photovoltaic solar cells has been and is still being further accelerated by the increased requirements of large spacecraft for missions in Earth orbit, for exploration and utilization of the Moon, and for exploration of the planets. In anticipation of these missions, under development are large lightweight deployable solar cell arrays with power levels up to 50 kW requiring about 5000 square feet of silicon solar cells for operation in 100 per cent sunlight.

In wide use today are N/P (negative on positive) silicon solar cells – the Apollo Telescope Mount will use cells 2 × 2 and 2 × 6 cm – which have superior radiation resistance, can cover large areas, provide better control over mechanical and electrical tolerances, and can withstand wide environmental extremes. Exposed to the space environment, these cells attain their highest efficiency, about 12 per cent of 0°C; at 25°C their efficiency drops to 10–11 per cent. The electrical interconnection of the solar cells contributes substantially to their high cost of fabrication. The cells are rather thick, 0·008–0·012 in, with cover slides of 0·003–0·030 in being utilized for specific missions.

In Earth orbit the efficiency of solar cell arrays is reduced and weight and cost penalties are introduced by the following factors:

● Cell equilibrium temperatures in excess of 30°C;
● Degradation of electrical contacts and inter-connections owing to thermal cycling, sensitivity to radiation and humidity before launch; and
● Cell degradation (owing to the space environment).

New processes such as the manufacture of solar cells from webbed dendrite silicon and from extrusion of a ribbon of silicon single crystals, now under study, are expected to increase the achievable cell size to 2 × 30 cm (or longer), thus resulting in long-range economic benefits, especially for the large solar cell arrays. Also under study for use in space are lithium-doped silicon solar cells which could provide a fifty-fold improvement in radiation resistance over the conventional N/P silicon cell. A potential cost reduction may also be obtainable by using optical concentrators to focus solar radiation on an individual solar cell, thus reducing the number of cells. Finally, feasibility studies for a 250 sq ft roll-out solar cell array indicate that weights of 30 W/lb are within the state-of-the-art.

Beyond the state-of-the-art, lower weight and lower cost requirements point towards the development of thin-film cells. For example, it would be desirable to develop deployable arrays weighing less than one pound per kilowatt. The cadmium sulphide thin-film cell is designed to provide weight and cost reductions. Large-area polycrystalline films have been prepared (for example, 3 × 3 in cells deposited on a metallized plastic substrate) and should prove useful if two major limitations – instability and low efficiency (about 5 per cent) – can be overcome.

The most significant long-term opportunity is for a major advance in photovoltaic efficiency. Since the photovoltaic cell is a quantum device when operating at a temperature substantially less than the source, it is not limited by conventional heat engine thermodynamics. While the single-transition silicon solar cell is theoretically limited to efficiencies of 20–25 per cent, solar cells with much higher efficiencies are theoretically possible. For example, a multicellular device (two or more photovoltaic layers in a sandwich) could use wavelength bands where the materials have high quantum efficiencies and thereby considerably increase their overall efficiency and reduce the heat rejection requirement.

Organic compounds which show characteristic semiconductor properties, including the photovoltaic effect, have only recently been considered as possible energy conversion devices. The state-of-the-art of organic semiconductors, however, is quite primitive, with overall efficiencies only a fraction of a per cent. Whether such compounds will be able to achieve the desired efficiencies is still a matter of conjecture, although a very attractive one because of the potential weight and cost reductions.

Transmission to microwave generators

The electrical power produced through photovoltaic conversion will have to be gathered at the solar collector and transmitted to the microwave generators. The high power levels will require that the transmission line be superconducting to reduce weight and power losses. For example, to transmit 10 000 MW (20 kV at 5×10^5 A), a superconducting line consisting of two conductors of 2 in diameter, each suitably insulated, would be required. Multiple-staged refrigerators would provide the desired temperatures over the length of the transmission line. At the superconducting temperature, 1000 W of refrigeration capacity would be sufficient to cool the line and to absorb heat leaks at the cable ends.

Because the solar collector has to be pointed approximately at the Sun while the radiating aerial is beamed to a receiving aerial on Earth, relative motion between the solar collector and the aerial will have to be provided. Rotary joints at the warm end of the transmission line with low friction and capability to carry the power will have to be developed. One approach would be to use liquid metal as the electrical conductor between the stationary and movable parts of the transmission line. Because of the substantial power density, switching would have to be accomplished at the individual microwave generators rather than in the transmission line.

Microwave beams

Transfer of power over a microwave beam has been demonstrated experimentally and is being considered for long-distance transmission. This approach has a present advantage over the newer field of laser power generation and transfer because materials and components for high-power lasers are still in the research or developmental stage. In addition, a radio window extending over the range from 10 cm to 1 m makes absorption of microwaves in the frequency range from 1 GHz to 10G Hz by the atmosphere negligible. Although requiring smaller transmitter and receiver apertures compared to microwave systems, laser high-power transmission would suffer losses in passing through the atmosphere. Eventually, a combination of laser transmission to a stationary platform just above the atmosphere and microwave transmission to Earth might be considered.

An example of the present technology is microwave valves (amplitrons) capable of generating hundreds of kilowatts of continuous power for the dc-to-microwave conversion. Laboratory models have achieved 425-kW outputs in the 10-cm band at 75 per cent operating efficiency – operating efficiencies of 90 per cent should be possible. The principle behind the amplitron tube is the use of a continuous crossed-field interaction such as in a conventional magnetron oscillator. In addition to the magnetron's high efficiency and simple construction (Fig. 48), the amplitron can amplify over a broad frequency band. In a double-stage amplitron the cascade arrangement produces 425 kW of continuous wave power at a wavelength of 10 cm and radiates r.f. power into space directly through a radome vacuum window. This tube operates over a bandwidth of 50 MHz at nearly constant efficiency, and typically requires a power supply capable of providing 20 kV at 15–20 A. To minimize atmospheric absorption of microwaves, the optimum frequency range of an amplitron is 1 GHz ($\lambda = 30$ cm) to 10 GHz ($\lambda = 3$ cm).

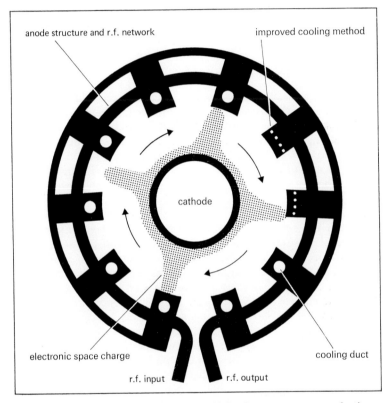

FIG. 48   Cross-section of amplitron for high microwave power production.

The transmission and conversion of microwave power to electricity has been demonstrated by tests in which a microwave beam from the Earth was used to power a helicopter a short distance away. Work is now under way to test similar transmissions of power to determine the feasibility of using them over greater distances, not only to power a helicopter but to guide and control it as well. The feasibility of microwave power transmissions between a space station and other satellites is also being investigated.

One of the problems in microwave generator systems is the heat

143

they generate and how to remove it. The amount of heat is directly a function of the efficiency of the generator system. The heat – on Earth it is removed by circulating water – could be radiated into space by means of heat pipes or a space radiator distributed over the structure of the microwave aerial. If one assumes that 10 per cent of the power generated in the amplifier is in the form of heat, perhaps as much as 1000 MW of residual heat would have to be removed. But if this heat is removed at a high enough temperature, it may be possible to use it – before radiating it into space – to run auxiliary power systems for space processing and manufacturing purposes.

Transmission of microwave beams

The microwave beam must be narrow so that the receiving aerial can intercept most of the beam. The width of the beam is largely a function of the diameter of the transmitting aperture and the wavelength of the radiation, but it also depends on the distribution of illumination intensity over the aperture. The beam is formed by uniform illumination of an aperture with a plane wave phase front. In the far-field regions, this wave pattern consists of a main lobe and several side lobes.

Gaussian illumination of the transmitting aperture produces a beam that has the same distribution pattern of energy along any plane parallel to that of the aperture regardless of the distance of that plane from the aperture. This type of illumination provides a Gaussian intensity distribution at any cross-section within the beam. The beam converges, reaches a minimum size and diverges again. The wavefronts are nearly spherical. The convergence of the beam is optimized when the waist of the beam is midway between the transmitting and receiving aperture and the diameter of the wave is equal to 70 per cent of the diameter at the two apertures.

One of the factors in the overall transmission efficiency is the efficiency of illumination of the aperture. The illumination efficiency of standard radar and communication apertures is usually about 70 per cent. Higher efficiencies can be achieved with horn-type or lens-type apertures but these are bulky. Phased arrays also provide illumination efficiency and are not necessarily bulky, but the methods of constructing them are complex. If the techniques for coupling microwave energy into the cavity from the microwave generator can be improved, the Gaussian beam transmitter may be highly efficient.

To produce the beam an array of amplitrons could be assembled and directed towards the aerial, whose curvature could be adjusted to obtain the desired wavefront shape. For example, if 1 MW amplitrons were used, 10000 amplitrons assembled into a 200 × 200 ft array would be required to produce the 10000 MW.

Microwave beam aerials

The aerial required to form the desired beam for the transmission of microwaves from geosynchronous orbits to Earth may be developed as an outgrowth of the various studies for large space-erectable communication aerials and radio telescopes. In 1965, the Space Science Board of the National Academy of Sciences recommended that radio telescopes with apertures of about 20 km be studied for possible application or adaptation to long-wave radio astronomy. To meet the limited payload volume, a rotationally deployed and stiffened paraboloidal dish 1500 m in diameter was analysed.* Although it had no bending stiffness of its own, its centrifugal forces enabled it to withstand perturbation loads and to maintain its shape. The dish consists of a structural support net with 250 meridionals and 40 circumferentials filled in with an aluminized polymer conductor grid. The whole structure spins around the axis of symmetry 1 revolution every 11 minutes to achieve the paraboloidal shape within 2 m over most of its surface. The mass of this structure is about 6000 lb. For microwave radiation where the wavelength would be between 10 cm and 1 m, this particular design is not adequate but its concept demonstrates one possible design approach.

Communications aerials are usually petal-type or folding-type thin-shell erectable-paraboloid or erectable-truss structures. A paraboloidal expandable truss aerial with an 80 per cent open metal-lized knit pre-coated weave material for the reflector would weigh about 0·1 lb per square foot of aperture. Its thermal distortion in orbit would be low and its natural frequencies – about 1–1·5 Hz in a diameter of 300 ft – would be compatible with the attitude control system. Using the present design, an aerial with a 1·5 × 1·5 mile aperture would weigh about 4 million lb. Considerable decrease in weight could be achieved if a design for the large-area aerial were developed.

Another type of communications aerial is a pressurized inflatable structure consisting of toroidal rings and stiffening members. Design studies have been carried out on an aluminized plastic reflector of 2000 ft diameter, to be deployed in a synchronous orbit. Based on the potential for technological advances in aerospace structures, a deployable structure for power transmission can be projected to weigh about 1 million lb.

Microwave aerial pointing

The microwave beam should lock on to a receiving aerial on Earth without straying more than 500 ft in any direction. To achieve this,

* Fager, J. A., *Large Space Erectable Communication Antennas*, IAF Paper SD9, 19th Congress of the Int. Astr. Fed., New York, October 1968.

the pointing accuracy will have to be less than one-half second of arc. As of now, the attitude control technique and pointing accuracy of the Apollo Telescope Mount to be flown in 1973 in a low-altitude Earth orbit over a 15-minute period of time will have an overall experiment package pointing uncertainty of approximately a rectangle 10 × 22 seconds of arc, corresponding to an error rectangle on Earth from a geosynchronous altitude of about 2 × 4 km. A system study for a manned orbiting telescope has shown that pointing accuracies of 0·01 second of arc could be achieved with a soft gimbal system.

The pointing accuracies required for the microwave beam aerials could be achieved once a structure compatible with guidance and control requirements and attitude control systems has been developed. An alternative approach to aerial pointing would be to develop electrical steering systems utilizing phased-array feeds.

Conversion to D.C. power

Microwave energy can be converted into d.c. power at acceptable efficiencies by point contact semiconductor diodes and thermionic diodes. These devices can operate continuously, provide an output impedance compatible with that of electric motors and handle high power. Although a single diode can only handle up to about 100 mW, many of them can be easily combined to form modules which can handle a considerable amount of microwave power. Standard subminiature diodes can be mounted in series parallel to form a single-phased full-wave rectifier. Schottky barrier diodes, which operate with efficiencies of 80 per cent and can handle a 200 mW output, can handle about 3 kW per lb of array.

Great improvements have been made in the 'rectenna' (a device that collects as well as rectifies the microwave energy by combining the receiving aerial (antenna) with solid-state rectifier diodes in such a manner that the energy is captured and rectified at efficiencies of 75 per cent or better). Theoretically, diode efficiencies greater than 90 per cent could be achieved. Rectennas have large, comparatively non-directive apertures that do not have to be pointed accurately in the direction of the transmitter. Furthermore, their residual heat can be dissipated by convective cooling. The rectenna's directivity is controlled by the directivity of the smallest aperture into which the rectenna is divided. In the rectenna shown in Fig. 49, the smallest aperture size corresponds to that of the half-wave dipole, a dipole whose directivity is not very great.

The rectenna's overall efficiency is determined by the product of the efficiency of its diodes and the collection efficiency of its aperture. In general, the aperture should appear as an emittance plane which is matched to the incident microwave illumination. If this match is perfect there should be no reflection, and the collection efficiency should be 100 per cent. Such efficiency can be approached by the use of a reflecting plane one-quarter wavelength behind the rectenna face, proper design of the rectenna face and best choice of d.c. load. The most difficult task in designing the structure is to prevent or minimize the generation of harmonics. With additional development, an overall efficiency of up to 69 per cent may be possible.

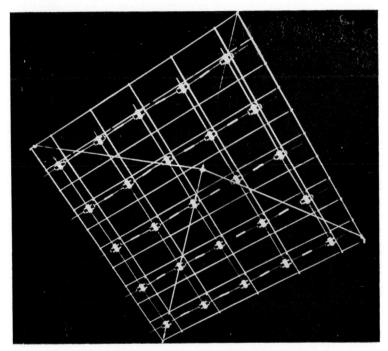

FIG. 49   Section of silicon rectifier diode aerial (rectenna).

System considerations

First, let us consider the payload. A reusable space shuttle is expected to result in significantly reduced costs of orbiting payloads in the 1975–85 time period. As envisaged, the space shuttle will be a fully recoverable two-stage vehicle consisting of an upper stage orbiter and a basic booster capable of carrying payloads of at least 50 000 lb. Current costs of orbital operations, excluding payload costs, range from $5000 to $10 000 per lb. The objective is to lower the cost per pound of payload placed in low Earth orbit to about $50 per lb and to lower the cost for synchronous orbit to about $500 per lb. A further reduction in cost is projected when a reusable nuclear stage for transporting men, spacecraft and supplies between low Earth orbit and geosynchronous orbit becomes available.

Satellite station assembly

The first metal-joining experiments with electron beam welding, consumable electrode welding and arc plasma welding using aluminium, steel and titanium were carried out in an experimental chamber by the Russian cosmonauts in one of the Soviet Union's three Soyuz-spacecraft during the flight on 20 October 1969. Similar metal-joining experiments are being planned for the orbital workshop because their perfection will be essential to the assembly of modular space stations envisioned for future missions (see Introduction).

These developments indicate that the capabilities for producing large structures in space will be available during the next decade. The experience gained in the assembly of such large structures by human operators – subsequently with the help of automated tele-operators – will be necessary if satellite solar power stations are to become a reality.

Orbit location

A geosynchronous orbit at a distance of 22 300 miles parallel to the Earth equatorial plane would make the satellite stationary with respect to any point on Earth, and make it subtend a 17-degree view angle of the Earth. The satellite would pass through the Earth shadow for about one hour each day for a total of 25 days preceding and following equinox. Thus, a system would require at least two satellites, each positioned so that at least one would be illuminated by the Sun at all times. The satellites would have to be about 21 degrees out of phase so that both would have a direct line of sight to the same point on Earth. A network of satellite stations would provide the most effective systems operation by permitting the supply of widely dispersed points on Earth either continuously or as required to meet peak power demands.

Microwave beam interactions

The microwave beam wavelength and frequency can be selected to minimize, if not totally eliminate, atmospheric absorption. Attenuation or absorption of millimetric waves propagating through the atmosphere is caused primarily by gas molecules with electric or magnetic dipole moments and by water formations such as clouds, fog and rain.

148

Absorption occurs at frequencies beyond 15 GHz, primarily due to molecular resonances of water vapour (22 GHz, 183 GHz), oxygen (line complex around 60 GHz, single line at 119 GHz), by a residual contribution of the entire water vapour and oxygen spectra extending into the far infrared and possibly to pressure-induced polarization in water vapour. Scattering and dielectric losses in the presence of condensed water above 15 GHz lead to attenuation, phase dispersion and atmospheric noise. The choice of microwave frequencies between 1 GHz and 10 GHz, therefore, could obviate the major causes of attenuation and absorption of millimetric waves.

The microwave beam power density would be reduced to as low a level as possible consistent with the desire to limit possible radiological health hazards. At levels such as 0.01 W/cm^2 the microwaves in passing through the upper atmosphere might cause a voltage gradient of about 2 V/cm, at which gradient ionization of the atmosphere would be very unlikely. The specific levels to be set for these voltage gradients will depend on an increased understanding of non-linear effects in the ionosphere such as electromagnetic drift caused by activity on the Sun and thermal and gravitational atmospheric tides, to ensure that losses will not occur when the beam passes through the upper layers of the atmosphere.

Microwaves have been recognized as a potential health hazard for more than a quarter of a century, but at a low power density of 0.01 W/cm^2 (one-tenth of the level of solar radiation received on Earth), the microwave beam would have destructive effects on neither objects nor living tissues exposed to it for a short period. In the United States, the allowable radiation energy absorption is 0.01 W/cm^2 for periods in excess of one-tenth of an hour; in Russia, permissible levels have been set at 0.001 W/cm^2. (Only recently have probes which could adequately measure microwave radiation at such low levels been designed.) Exposure at a level of 0.01 W/cm^2 would only occur in the beam proper, but with suitable safety devices and regulations, entry of living beings or objects into the beam could be avoided to minimize any possible long-term biological effects.

Satellite and earth installation dimensions

The power to be provided by a satellite station would probably first be used to meet peak power demands and to supplement power to existing grid systems. As the system is developed, power would be provided to specific areas where its use would be preferable over other forms of power generation. For example, power of about 10000 MW received on Earth would meet the needs of a city the size of New York. To absorb this power on Earth would require an area of about 6 × 6 miles. If we assume a system efficiency of 10

L*

per cent, a satellite solar collector, based on state-of-the-art components, would occupy about 5 × 5 miles. To radiate 0·01 W/cm² the microwave aerial would be about 1·5 miles in diameter (Fig. 44).

Weight and cost projections

Detailed design concepts for a satellite solar power station have not yet evolved to the point that cost and weight trade-offs among components can be made. However, the first step towards this goal can be taken by projecting from the present state-of-the-art the direction that future developments may or will have to take. The projected weights for the major components for the satellite portions of a system designed to provide about 10 000 MW of power on Earth are as follows:

Solar collector	$2·5 \times 10^6$ lb
Aerial	1×10^6 lb
Microwave equipment	1×10^6 lb
	$4·5 \times 10^6$ lb

These projections were based on the following assumptions:
- The solar collector weight is about an order of magnitude decrease in the weight of state-of-the art components. The basis for the projection is a solar cell weight of 10^5 lb per square mile. The use of organic photoconductors could reduce this weight by at least one-half.
- Experimental and design studies indicate that at high power levels, microwave generators will weigh about 0·05 lb/kW. The cooling equipment is projected to weigh 0·5 lb/kW.
- The microwave aerial weight is based on a weight reduction of about an order of magnitude from present designs such as the expandable truss.

The projected costs for the system are as follows:

Insertion into synchronous orbit	$ 45/kW
Earth receiving aerial	$100/kW
Solar collector	$300/kW
Microwave generation	$ 50/kW
	$495/kW

These projections were based on the following assumptions:
- Payload insertions costs will be about $100 per lb.
- Near-term projections indicate that large silicon solar cells would cost about $1000 per kW when a market would justify considerable development costs for manufacturing improvements. Thin-film cells could be produced at somewhat lower

costs compared to single-crystal cells. (This is the greatest area of uncertainty. In present spacecraft, the solar panels cost $100 000 per kW. The basic cost of silicon is about $100 per lb or about $30 per kW.)

- Near-term cost projections for microwave generators are about $10 per kW; $40 per kW is allocated for the cooling system.
- The Earth-receiving aerial cost is based on the already achieved rectification of microwaves with three pounds of solid-state rectifiers per kW and the expectation that the quantity of materials required will be reduced severalfold.

The projected system cost of $500 per kW is about twice that of present power generation costs for fossil fuel or nuclear energy sources. The assumption of 10 per cent system efficiency is a conservative one and has been chosen to illustrate that no significant technical advances are required to achieve it. It is highly likely that the system in its final form would have a substantially higher efficiency and thereby achieve a competitive advantage compared to other power generation systems.

Authors

JEAN A. VANDENKERCKHOVE (*Technology of Earth Satellites*), aged 43, has been Assistant Director in charge of the Space Applications Division at ESRO Headquarters in Paris since 1969. He graduated as a mechanical and electrical engineer (1952) and Doctor of Applied Research (1962) from Brussels University. He was also graduated M.Sc. in Mechanical Engineering from the California Institute of Technology in 1954. Dr Vandenkerckhove began his career as Assistant attached to Prof. Von Karman in Brussels University, where he has lectured on rocketry and space technology since 1960. Until 1962 he was a consultant for Belgian and US industry, mainly on rocket motors and solid propellents. During the period 1961–62 he was technical adviser to the Belgian delegation in the negotiations which led to the creation of ESRO and ELDO. In April 1962 he joined the preparatory commission for ESRO in Paris and in 1964 was transferred to the European Space Technology Centre in the Netherlands. Dr Vandenkerckhove was Project Manager for the HEOS-1 satellite, launched successfully in 1968. He is the author of more than 50 articles and publications and co-author of *Rocket Propulsion*, published in French, English, German and Russian.

JOHN KENNETH SAVILE JOWETT (*Communications Satellites*), aged 55, is Deputy Director Engineering in P.O. Telecommunications Headquarters, responsible for developments in space communications. He has spent almost all his career with the British Post Office in the field of radio-communications. His activities in this specialized field during the past 10 years have included representation of Britain at international technical conferences and, in particular, attendance at some 30 meetings of the Technical Sub-Committee of the INTEL-SAT consortium. Earlier in his career Mr Jowett was associated with the planning and development of overseas HF radio-links, of VHF links to the islands of Britain and of SHF radio-relay links for the national wide-band telecommunications network. He has written a number of technical articles on radio-wave propagation, broadcasting service planning and satellite communications. At his home in Cheam much of his time is filled with duties as secretary of a local church but he still maintains some time for his garden and for the occasional game of cricket.

BASIL JOHN MASON (*Meteorological Satellites*) has been Director-General of the UK Meteorological Office since 1965, having previously been Professor of Cloud Physics at Imperial College, London. Dr Mason, who is 48, graduated with first-class honours in Physics at London University in 1947, and was awarded the degree of Doctor of Science in 1956. His research has been mainly with the mechanism of weather – formation of clouds, rain, snow and lightning; subjects on which he has published a major monograph *The Physics of Clouds* (Clarendon Press 1957) and a more popular text *Clouds, Rain and Rain-making* (Cambridge University Press 1962). He was elected into the Royal Society in 1965, and was recently President of the Royal Meteorological Society and Honorary General Secretary of the British Association for the Advancement of Science. Dr Mason is the Permanent Representative of the UK at the World Meteorological Organization and a Member of its Executive Committee. His hobbies are foreign travel and music.

WATSON LAING (*Earth Resources Satellites*), 33, is Design Project Leader, Future Projects Studies, of Hawker Siddeley Dynamics' Space Division at Stevenage. He has an honours degree in physics from St Andrew's University and is a Fellow of the British Inter-planetary Society; also a member of the Institute of Physics, the British Computer Society and the American Society of Photo-grammetry. Mr Laing has led several studies in Earth resources, covering technology, economics, and interpretation and the design of instruments. He also worked on the feasibility studies and project definition of the British X4 satellite, for which he was Experiment Controller. He has also recently led a study of a stellar astronomy module compatible with manned space stations as part of European post-Apollo activities, and is currently working on Earth resources studies and on development cost estimating techniques. Mr Laing lives in Stevenage with his wife and daughter.

ROY E. ANDERSON (*Navigation satellites*), aged 53, is a consulting engineer at General Electric's Research and Development Center in Schenectady, New York. The Company recently appointed him as a Coolidge Fellow in recognition of his engineering achievements. Mr Anderson has generated concepts and provided technical leadership in studies and experiments aimed towards satellite systems that will provide automatic position fixing and communications for traffic control of mobile craft over large regions of the Earth. He joined the General Electric in 1947, and has contributed to developments in radar, radio direction finders, telemetry, encoding and recording systems, and industrial electronic devices. Previously he was an instructor in physics at Augustana College and served two years as a US Navy electronics officer. Mr Anderson holds a BA degree in physics from Augustana College and a MS degree in electrical engineering from Union College. He is married with four children.

PETER STUBBS (*Research Satellites*) is Deputy Editor (Science) of the weekly magazine *New Scientist*. He was educated at Christ's Hospital and Nottingham University. After graduating, he spent eight years, from 1952 to 1960, at Manchester University and Imperial College, London, on research in palaeomagnetism and its relevance to continental drift and polar wandering theories. Dr Stubbs joined *New Scientist's* staff in 1960 and now writes on a wide variety of topics in the physical sciences. His special interests lie with the Earth sciences. He is 43, and likes good music and mountaineering.

NEVILLE BROWN (*Military Satellites*), aged 39, is a Senior Lecturer in International Politics at the University of Birmingham, and writes regularly on defence topics for *Le Monde Diplomatique* and *New Middle East*. Formerly a Lieutenant in the Meteorological Branch of the Fleet Air Arm, he has also worked on the staffs of the Royal Military Academy, Sandhurst, and of the Institute of Strategic Studies. Mr Brown, a former defence correspondent of *New Statesman* and *New Scientist*, is author of several books on strategic studies and contemporary history, among them *Arms without Empire* (1967), *British Arms and Strategy: 1970–80* (1969), and he contributed Volume 3, 1945–1968 to *The History of the World in the 20th Century* (1970). He lives at Watlington, Oxon.

PETER E. GLASER (*Advanced Satellite Concepts*) is head of the Engineering Sciences Section at Arthur D. Little, Inc., Cambridge, Massachusetts. Since joining the staff in 1955, he has directed research on methods of generating high temperatures and on methods of measuring the Earth–Moon distance with laser ranging retroreflectors (placed on the Moon during the Apollo 11 mission) and the heat flow from the lunar surface with probes (installed during the Apollo 15 mission). Dr Glaser is Past President of the Solar Energy Society, a member of committees of the National Academy of Sciences and Vice-President of the Heat Transfer Commission of the International Institute of Refrigeration. He has numerous publications, books and patents in the fields of his activities. Dr Glaser lives in Lexington with his wife and three children. He has an interest in the archaeology of southern Arabia and enjoys gardening and recreational sports.

154

DAVID FISHLOCK (Editor) also edited the companion volumes *A Guide to the Laser* (1967) and *A Guide to Superconductivity* (1969). He is 38 and for four years past has been Science Editor of *The Financial Times*. He is author of several other books, among them *Man Modified* (1969) and *The New Materials* (1967).

Subject Index

157

ICBMs, 126
Infrared Interferometer Spectrometer (IRIS), 57
Infrared linescan imagery, 76
Infrared surveillance, 54, 124, 125
Intelsat, 23, 25, 26, 36, 39, 40-42, 44
Intelsat 3, 25, 30, 31, 33, 37, 38
Intelsat 4, 32, 37, 38
International Radio Consultative Committee (CCIR), 105
Interplanetary Monitoring Platform (IMP), 119
Interrogation, Recording and Location System (IRLS), 64, 65, 94

Laser instruments, 75
Launch vehicles, 2-6
Loran navigational system, 68, 98
Low altitude surveillance platform, 126
Lubricants, solid, x
Lunar, 9, 115
Lunar Orbiters, 113, 115

Manned Orbiting Laboratory (MOL), 132, 133
Mariners, 113, 115, 118
Marshall Space Flight Center, viii
Masses of *Intelsat* sub-systems, 37-40
Mercury spacecraft, 79
Meteor, German research ship, 101
Meteor weather satellites, 46
Meteorological Satellites, 46-68
Microwave beams, 142-146, 148, 149
Microwave instruments, 75
Microwave regions, 75
Military Uses of Satellites, 121-133
Missile Defense Alarm System (MIDAS), 124, 125
Missile policies, 126, 127
Molnya, 13, 26, 30
MRBMs, 127
Multispectral scanner, 83

National Aeronautics and Space Administration (NASA) vii, viii, 3, 23, 64, 66, 79, 85, 94, 96, 98, 100
Navigation vehicles used in passive tests, 101, 102
Navigational Satellites, 92-105
NAVSTAR system, 104
Nimbus, 46, 50, 51, 54, 87, 94
Nimbus 3, 46, 47, 57, 79
Nimbus 4, 57, 63
Noise budget of *Intelsat* 3, 32

Noise, radio, 28, 31, 32, 33
North American Rockwell, x

OMEGA navigation system, 65, 67, 68
Omega Position Locating Equipment (OPLE), 65, 66, 98
Optical laser radar (LIDAR), 75
Orbit choices, 10-14, 118-119, 148
Orbita system, 26, 43
Orbital bombs, 129, 130
Orbital periods, 24
Orbiting Astronomical Observatory 2 (OAO 2), 110, 118
Orbiting Geophysical Observatory (OGO), 87, 119
Orbiting Solar Observatories, 109, 110, 119

Pegasus Experiment, 116
Philco-Ford, 97
Photographic swath widths, 78
Planck's Law, 72
Planetary spectaculars, 111
Political and legal aspects, 89, 90
Position Location and Aircraft Communication Equipment (PLACE), 104

Quartz emission energy distribution, 73

Radiated energy measurements, 53-61
Radio location-active systems, 93-98
Radio location-passive systems, 98-105
RAE 1, 117
Rectenna, 146, 147
Relay satellite, 23
Remote sensing, 72-78, 84
Research Satellites, 106-120
Restraints, vehicle, 7-9
Rocks spectral signature, 74

Salyut, vii
Samos, 69, 125
Satellite design, 15, 16
Satellite interception programme (SAINT), 128, 129
Satellite's orbital plane, 9
Satellite station assembly, 148
Satellites, *see*
    Advanced Concept Satellites
    Communication Satellites
    Earth Resource Satellites
    Earth Satellite Technology
    Meteorology Satellites
    Military Uses of Satellites